Pollutant Transfer and Transport in the Sea

Volume I

Editor

Gunnar Kullenberg

Professor of Physical Oceanography
Institute of Physical Oceanography
University of Copenhagen
Copenhagen, Denmark

CRC Press, Inc.
Boca Raton, Florida

Library of Congress Cataloging in Publication Data
Main entry under title:

Pollutant transfer and transport in the sea.

   Includes bibliographical references and index.
   1. Marine pollution. 2. Ocean circulation.
I. Kullenberg, Gunnar.
GC1085.P63   628.1′686162   81-4500
ISBN 0-8493-5601-6 (v. 1)   AACR2
ISBN 0-8493-5602-4 (v. 2)

   Direct all inquiries to CRC Press, Inc., 2000 N.W. 24th Street, Boca Raton, Florida, 33431.

© 1982 by CRC Press, Inc.

International Standard Book Number 0-8493-5601-6 (Volume I)
International Standard Book Number 0-8493-5602-4 (Volume II)

Library of Congress Card Number 81-4500
Printed in the United States

PREFACE

The distribution of pollutants in the sea is influenced by physical, chemical, and biological transfer processes. These processes interact, and in order to understand the pollution problems and implications for the marine environment it is necessary to take this into account. The aim of this book is therefore to give a presentation of various physical, chemical, and biological transfer processes which includes an account of our present understanding of relevant aspects of the marine systems, discussions of the state of the art as regards parametrization and modeling of transfer processes, and an account of applications to some specific examples.

The subject areas are covered by different authors and the editor has not attempted to really integrate the various chapters. Rather, the aim has been to cover those subject areas which we now consider essential for the transfer and transport of pollutants in the marine environment, obviously without claiming complete coverage. The basic message is that a proper understanding of the transfer of pollutants in the marine environment requires an interdisciplinary approach and due consideration of physics, chemistry, biology, and geology.

The individual chapters by and large stand on their own. A list of notations has been included in the individual chapters when required, but it should be noted that all notations have not been unified. That was simply not feasible.

The first two chapters of Volume I deal with physical processes in the ocean and predictive modeling of transfer of pollutants. Chapter 1 is an attempt to present recent ideas and observations of the physical conditions in the ocean thought to be relevant to the pollution problem in general. Chapter 3 gives a very brief account of techniques commonly used to investigate experimentally the physical spreading of pollutants on small to mesoscales. Chapter 4 gives an account of the air-sea exchange of pollutants, where much progress has been made in recent years.

Processes and transfer related to biological conditions, suspended matter, and sediments are considered in Volume II, Chapters 1, 2, and 3. In all these subject areas the research is moving fast and several new results are presented here.

In Chapter 4 estuaries and fjords are discussed separately since these coastal zones are considered to be of special interest. Likewise in Chapter 5 many aspects of the spreading of oil from oil spills are considered, including processes and modeling with a discussion of findings from several recent spills, since oil spills constitute a serious problem in many areas. It is hoped that this publication will stimulate discussions and interdisciplinary research relevant to pollution problems as well as serve as an educational reference book.

G. Kullenberg

THE EDITOR

Gunnar E.B. Kullenberg received a Ph.D. in Oceanography from Göteborg University, Göteborg, Sweden, in 1967.

Since completion of his doctoral work, he has held positions as Associate Professor of Oceanography at the Institute of Physical Oceanography, University of Copenhagen, 1968-1977; Professor of Oceanography at Göteborg University, 1977-1979; and his current position since 1979, Professor of Physical Oceanography, Institute of Physical Oceanography, University of Copenhagen.

Dr. Kullenberg's main research interests include turbulent mixing, air-sea interaction, optical oceanography, problems related to marine pollution, and coupling between physical and biological processes.

His participation in several international programs has led to work not only in northern European waters but also in the Mediterranean, central and eastern subtropical Atlantic, Southern, and Pacific Oceans.

Approximately 50 papers have been published since 1968, including review papers and presentations at several international symposia and workshops.

Dr. Kullenberg is engaged in the following international organizations: ICES (International Council for the Exploration of the Sea), SCOR, IAPSO (International Association for the Physical Sciences of the Ocean), GESAMP (the U.N. Technical Agencies Joint Group of Experts on Scientific Aspects of Marine Pollution), and in the work of some U.N. Agencies directly. He is chairman of the joint ICES and SCOR WG on the study of the pollution of the Baltic, chairman of ICES Advisory Council on Marine Pollution, and chairman of the GESAMP WG on the health of the ocean. He is engaged in the IAPSO Committee on dispersion problems and inthe WG on ocean optics. In addition to these international engagements, he is involved in the work of Danish and Swedish national committees including the subcommittee on earth sciences of the Swedish Natural Science Council.

CONTRIBUTORS

R. Chester, Ph.D.
Reader in Oceanography
Department of Oceanography
University of Liverpool
Liverpool, England

E. K. Duursma, D. Sc.
Director
Delta Institute For Hydrobiological
  Research
Yerseke, The Netherlands

Scott W. Fowler, Ph.D.
Head, Biology Section
International Laboratory of Marine
  Radioactivity/IAEA
Musée Oceanographique
Principality of Monaco

Herman G. Gade, D. Philos.
Head, Department of Oceanography
Geophysical Institute
University of Bergen
Bergen, Norway

Gunnar Kullenberg, D.Sc.
Professor of Physical Oceanography
Institute of Physical Oceanography
University of Copenhagen
Copenhagen, Denmark

Stephen P. Murray, Ph.D.
Assistant Director and Professor
Coastal Studies Institute
Louisiana State University
Baton Rouge, Louisiana

Maarten Smies
Environmental Toxicologist
Shell International Research
  Maatschappij BV
Group Toxicology Division
The Hague, The Netherlands

Dr. G. C. van Dam
Head of Physics Division
State Public Works
Directorate of Water Management and
  Research
The Hague, The Netherlands

Michael Waldichuk, Ph.D.
Senior Scientist
Pacific Environment Institute
Fisheries and Marine Service
Canada Department of Fisheries and
  the Environment
West Vancouver, B.C.
Canada

TABLE OF CONTENTS

Volume I

Volume II

Chapter 1

PHYSICAL PROCESSES

Gunnar Kullenberg

TABLE OF CONTENTS

I. INTRODUCTION

The physical transport and transfer of pollutants is determined by the motion in the sea. The motion covers a very large range of scales, from the ocean-wide circulation to molecular motion, distributed over a more or less continuous spectrum (see Figure 1). The spreading, or dispersion, of a passive contaminant is governed by two classes of processes; namely, advection and mixing, sometimes also called diffusion. The advection, i.e., the average velocity over some space or time scale, is caused by relatively large-scale water movements transporting the given property and thus effecting a local change in concentration. The mixing is governed by comparatively small-scale random movements which give rise to a local exchange of the given property without causing any net transport of water. The combined effect of mixing (turbulent diffusion) and advection is here called dispersion.

The theoretical treatment of mixing attempts to relate all the mixing except the part caused by molecular movements (molecular diffusion) to the field of motion in the sea. In order to be able to do this, a detailed knowledge of the field of motion over the complete range of scales is required. At the present this knowledge cannot be obtained practically by means of observations at all scales, and our understanding of the dynamics of the motion is not good enough to permit a deduction of the whole field of motion from observations covering a limited range of scales. Usually it is necessary to rely on averaging procedures whereby a mean current field is defined by the averaging time or space scale which is appropriate for the particular problem.

The definition of a mean and a fluctuating part of the motion requires a separation between the mean and the fluctuating part which will depend to some degree upon the question at hand. In mixing considerations one is often concerned with the effects of small-scale motion, and the fluctuating part may be defined by the scales of motion which contribute significantly to the mean square shear or dissipation scales.

Depending upon the choice of averaging scale, different components of motion (e.g., tidal, meteorological, seasonal) are included in the mean current. The mean current causes advection of the contaminated fluid volume as a whole. The smaller scales perturb the mean motion in a more or less random way. There are different ways of expressing the effect of those scales on the spreading of the contaminant, i.e., to parameterize the mixing or diffusion process. The most common approach is to use effective turbulent exchange coefficients (or eddy diffusion coefficients) analogous to the molecular diffusion coefficients. This is a very crude approach which, however, often has to be relied upon for practical reasons. An alternative approach is to use more sophisticated models of turbulence to close the system of equations.[1,2] This requires an insight into the characteristics of the fluctuating random motion in the sea which is only gradually being obtained.

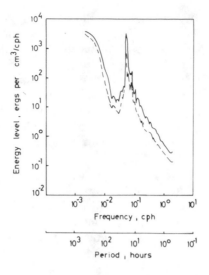

A

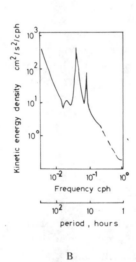

B

FIGURE 1.   (A) Distributions of oceanic motion as function of frequency (or period). Eulerian horizontal kinetic energy full drawn at 511 m depth, dashed at 1013 m depth in the western North Atlantic. (From Fofonoff, N. P., Ref. 68—72, Geophysical Fluid Dynamics, Program, Course Lectures and Seminars, Vol. I, Woods Hole Oceanographic Institution, Woods Hole, Mass., 1968. With permission.) (B) Example of kinetic energy density spectrum from the central North Atlantic. (From Gould, W. J., Schmitz, W. J., Jr., and Wunsch, C., *Deep Sea Res.,* 21, 911, 1974. With permission.)

The diffusion generated by fluctuating random motion is generally larger than the molecular diffusion. Therefore the effective turbulent exchange coefficients are not directly influenced by the particular value of the molecular coefficients. This does not mean that the molecular diffusion can be neglected since ultimately it must be responsible for the dissipation. The action of the random motion is to deform or convolute the contaminated fluid volume thereby increasing both the concentration gradients and the surface areas over which the smaller scales of motion can act, down to molecular diffusion. The effect of the advective motion may be called stirring and the effect of conduction or diffusion, which decreases the mean value of the concentration gradient, may be called mixing.[3]

The type of analysis to be applied will vary depending upon the ratio of the scale of the random motion and the linear dimension of the contaminated volume. When this ratio is small, a Eulerian type of analysis is appropriate for determining the probability of finding a given particle inside a given volume at a given time. In this case the center of mass of the diffusing fluid volume is practically stationary. When the ratio is large, a Lagrangian type of analysis is appropriate for determining the characteristics of a moving, contaminated volume. In this case the motion in different parts of the contaminated volume is small relative to the translation of the whole contaminated volume. Then the distortion of the fluid volume is the significant factor, and it is thus necessary to follow the motion of the center of mass. The Lagrangian type of analysis implies the study of velocity correlations between two or more particles, whereas in the Eulerian analysis the motion of only one particle is considered. The type of analysis will also depend upon the particular problem at hand: in the case of a continuous injection at a fixed point a Eulerian type of analysis is normally used. The effects of advection and diffusion on the concentration are thereby included and the dispersion is in this case often referred to as absolute dispersion. In the case of an almost instantaneous source the Lagrangian analysis is normally used and the so-called relative dispersion is determined.

The study of turbulent diffusion aims at a description of the mixing of a contaminant in terms of the statistical characteristics of the turbulence. Progress is hampered by lack of data on the characteristics of the motion in the sea and by the difficulty of obtaining well-defined experimental conditions making it possible to distinguish important processes. A discussion of the dispersion in the sea should start with a summary of what is known about the motion in the sea and the factors influencing the motion. The main physical characteristics of interest in the present context are:

1. The spectrum of the motion covers a large range of scales. The scale of the fluctuating motion cannot in general be regarded as small relative to the characteristic length scale of the problem at hand. The larger scales of motion should generally be considered as statistically nonstationary and nonhomogeneous.
2. The water column is usually stably stratified, with a two-component system. The small-scale, mechanically forced vertical motion tends to be suppressed by the stratification. On basin-wide scales, however, there will be generated a vertical circulation which is essentially driven by the temperature and salinity differences. This type of motion tends to occur along density surfaces, which are inclined to the horizontal. Vertical transport can also be generated by double-diffusion, due to the difference between the molecular diffusion for heat and salt, in areas where one component (heat or salt) counteracts the other stratifying component.
3. The energy supply to the motion occurs at a number of frequencies and scales: the global wind systems, the short period wind fluctuations, the tidal frequencies, the surface gravity waves, and through thermo-haline processes. The motion in the deep sea is driven both by wind effects and by thermal forcing.

4.	The ocean is generally inhomogeneous. Transient and semipermanent fronts are common, as are well-defined eddies on various scales. The motion often follows isopycnals,* making the partition in horizontal and vertical components rather artificial or misleading. Over large spectral intervals the motion is highly intermittent, and the fluctuating velocities are larger than the mean.

The main environmental influence arises from the baroclinicity of the ocean and from usually stable stratification. The motion is called baroclinic when it varies with depth, i.e., has a vertical shear. Motion independent of depth is called barotropic, which means adjusted to the pressure field. The terms baroclinic, thermoclinic, and haloclinic refer to cases where the fields in question are inclined relative to the pressure, temperature, and salinity field, respectively.[4]

The stable density stratification implies that the turbulence in the ocean is weak. Wave motion plays an important role. Two extreme cases may be visualized. In cases of near-neutral density stratification the motion may be turbulent and characterized by high rates of energy dissipation and transport of momentum and matter. In cases of very stable stratification the turbulence will be suppressed and a field of internal waves will develop. The momentum transport may still be efficient but the transport of matter is suppressed, being mainly coupled to the breaking of the waves. The two types of flow are distinctly different, but are probably often simultaneously present in the sea.

The surface layer of the sea is directly influenced by the atmosphere. Above a certain wind speed the wind-induced motion will generate a surface mixed layer (Figure 2). The thickness of this layer depends primarily on the strength and duration of the wind and the insolation but is also influenced by the rotation of the earth. The layer is often separated from the deep water by a transition zone, a pycnocline layer (thermo-, halocline, or both), where the density increases relatively rapidly towards the density of the deep water. In some cases opposing density variations due to heat and salt may nearly compensate so that the pycnocline is weak although the thermo- and halocline are well developed.

In the open ocean the warm water sphere is separated from the deep cold ocean water by a permanent thermocline called the main thermocline. This lies at depths around 300 m in the equatorial region, deepening to 600 to 800 m in the subtropical areas and then gradually shallowing and reaching the surface at latitudes around 50°S and 55°N (see also Figure 3).

Latitudinal variations of temperature, salinity, and incoming solar radiation are of great importance for the conditions.[4a] Annual temperature variations in open ocean surface waters in arctic regions are about 2°C. In temperate latitudes, around 40°N and 40°S, they are up to 10 and 5°C in the northern and southern hemisphere, respectively, and less than 2°C around the equator. Around 10° latitude the seasonal variations are 2 to 3°C. The total temperature range is about 30°C, from tropics to arctics. Across frontal zones the change of temperature over some tens of kilometers can be of the order of 10°C. It should be noted that the seasonal temperature variations in coastal waters normally are considerably larger than those in the open sea.

The mean surface salinity in the open ocean also shows a very distinct latitudinal variation with maxima around 30°N and 30°S, decreasing towards higher latitudes and with a marked minimum around 5°N. The total range is about 33.5 to 37 o/oo. In coastal waters the range can be considerably larger.

---

*	Isopycnal or isopycnic surface = surface of equal density; isopycnals = lines of equal density in a section.

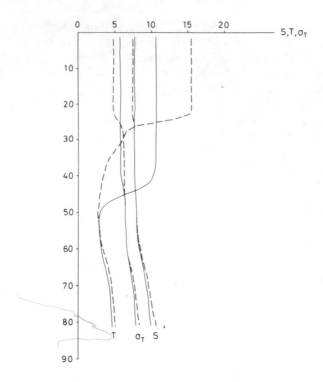

FIGURE 2.   Observed continuous profiles of temperature T, salinity, S, and density σ_t in the Baltic,
56°05′N, 18°23′E, before (dashed) and after (full
drawn) severe storm.

The total annual radiation reaching the sea surface does not show large differences.
The seasonal variability is large in temperate latitudes.

The tidal range shows a marked variation with latitude which is very important for
the conditions in coastal waters. There is generally low tidal range in high latitudes
and about as low, or around 1 m, around 20°N and 30°S. Maximum tidal range is
found in temperate zones, around 50 to 60°N and 45 to 55°S, with a range of 3 to 4
m. Around the equator there is a secondary maximum with a range around 2 m.

Finally it should be noted that studies of the dispersion of substances should take
into account also the lifetime of the substance in question, since the lifetime will essentially determine the characteristic timescales of physical processes effectuating the dispersion. Knowledge of the dispersion of a short-lived tracer cannot necessarily be applied to the dispersion of a long-lived tracer.

Following this outline of some basic conditions, a more detailed discussion will be
given of some recent observations regarding the characteristics of the motion and the
distribution of properties in the ocean.

II. CHARACTERISTICS OF THE MOTION

Since around 1965 a large number of current measurements have been carried out
in various parts of the ocean by means of moored automatic recording instruments.
The energy density spectra (see Figure 1) generally show peaks at the tidal (diurnal and
semidiurnal) and inertial periods, but with the bulk of the energy contained in longer
period motions.[5,6] The energy density of the short period fluctuations decreases rapidly
with increasing frequency. Webster[7] found an energy density decrease conforming with
a slope of −5/3 for periods less than 6 hr.

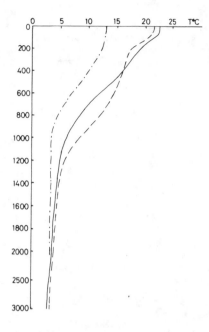

A

FIGURE 3.   (A) Temperature profiles at different latitudes in the North Atlantic. Full drawn, 20°N; dashed, 40°N; dotted 50°N. (B) Salinity profiles at different latitudes in the North Atlantic (same as in Figure A). (C) Profiles of salinity, temperature, and density, $\sigma_t = (\varrho-1) \cdot 10^3/\varrho$, where ϱ is density for the observed S and T and atmospheric pressure. (From Neumann, G. and Pierson, W. J., *Principles of Physical Oceanography,* 1966, p. 135. Reprinted by permission of Prentice-Hall, Inc., Englewood Cliffs, N.J.)

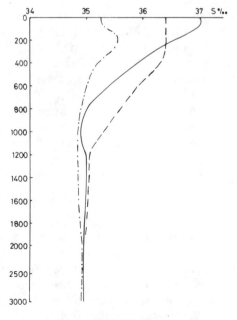

FIGURE 3B

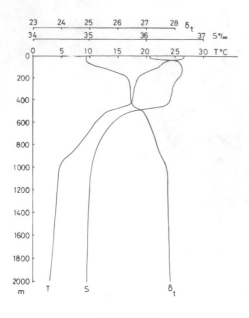

FIGURE 3C

A. Mesoscale Fluctuations

Already the early observations using neutrally buoyant floats showed the presence of long period fluctuations (referred to as eddies, which however, does not imply any closed path) with wavelengths of several hundred kilometers and periods of 50 to 100 days.[8] Further measurements extended in both time and space, using both neutrally buoyant floats and moored instruments, have clearly documented the presence of such eddies. These are referred to as mesoscale eddies, of typically of the order of 100 km wavelength and periods in the range 50 to 100 days.[9,10] Low frequency motion of other types have also been observed recently.[11,12]

Wyrtki, Magaard, and Hager[13] used observations of surface drift currents by ships to calculate the kinetic energy of the mean flow and the fluctuations. High energy density was found in the western boundary currents and in the equatorial current system whereas low values were typical of the subtropical gyres. The ratio between the eddy kinetic energy and the mean kinetic energy varied considerably, from 10 to 20 in the low-energy areas to 2 to 5 in the high-energy areas. This clearly shows the importance of the eddy or fluctuating motions in the surface layer.

Schmitz[14-16] has used subsurface current meter data to demonstrate the importance of the fluctuating kinetic energy for the layer from the main thermocline to the deep water, in the depth interval 600 to 4000 m. It was shown that the low-frequency kinetic energy per unit mass with periods larger than a day could vary by 2 orders of magnitude, in the range $(1—100) \cdot 10^{-4}$ m^{2}s^{-2}, at depths around 4000 m in the western North Atlantic, and that the energy was greatest in the vicinity of the Gulf Stream. The corresponding values for the surface layer found by Wyrtki et al.[13] were in the range $(400—600) 10^{-4}$ m^{2}s^{-2}.

Schmitz' findings suggest stronger eddy activity in connection with the Gulf Stream than in the subtropical gyre. Schmitz also showed that the energy-preserving kinetic energy frequency spectra were spatially inhomogeneous, both vertically and horizontally. At 28°N latitude at 4000 m depth the spectrum showed a well-defined temporal mesoscale, i.e., periods in the range of 50 to 100 days, whereas in the thermocline layer the spectrum showed increasing energy with decreasing frequency, the most pronounced signal there being in the range of periods from 100 to 1000 days. The spectra

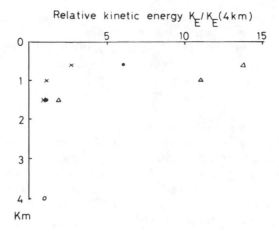

FIGURE 4.   Kinetic energy decrease with depth in central North Atlantic. (From Schmitz, W. J., Jr., *J. Mar. Res.*, 36, 295, 1978. With permission.)

from the thermocline layer in the Gulf Stream area, 38°N, and the spectrum from 4000 m depth at 28°N, 69.7°W (MODE Center) were, however, quite similar. In the Gulf Stream area the spectra from the thermocline layer and 4000 m depth showed a well-defined temporal mesoscale, peaking around periods of 50 days. Moving up the continental rise from the Gulf Stream[17] and the continental slope[14] there is a tendency for a shift in energy towards higher frequencies below the thermocline.

In the Gulf Stream eddy kinetic energy decreases considerably from the thermocline layer to the deep water, at 28°N, 69.7°W, and at the low-energy site 28°N, 55°W (Figure 4).[16] Normalizing the vertical profiles with the value at 4000 m depth Schmitz found the smallest variation within the Gulf Stream area and the maximal variation at the low-energy site in the subtropical gyre. There was also a difference in relative vertical distribution between different frequency bands. Periods less than 100 days, i.e., the mesoscale eddies, had essentially the same vertical distribution in the Gulf Stream area as in the MODE Center area. Periods larger than 100 days were more depth dependent at the latter position, whereas in the Gulf Stream area both ranges of time scale had very similar vertical energy distribution. In the thermocline region the eddy energy varied from about $10^{-3} m^2 s^{-2}$ at the low-energy site to about $2.9 \cdot 10^{-2} m^2 s^{-2}$ in the Gulf Stream area. Gould et al.[10] found that around 28°N, 69°W, 70% of the total energy was contained in the barotropic and the first three baroclinic modes, with more than half (54%) in the barotropic mode. Phillips[18] found from measurements with neutrally buoyant floats[19] (also in the western North Atlantic) that 78% of the total kinetic energy was contained in the barotropic mode.

The results presented by Schmitz[16] are very similar to other results obtained in other areas although the data material may be less complete. Eddy motion has also been observed in the eastern North Atlantic.[20] There is no strong reason to believe that eddy motion of the type discussed here should not be a general feature of the oceanic motion. The kinetic energy of fluctuating motion of this type is considerably larger than the kinetic energy of the mean motion. Clearly these results are of importance when considering the transfer of pollutants.

For the parameterization of the mixing by means of diffusion coefficients a gap in the spectrum of the motion is required, and the observations suggest the existence of such a gap. On the other hand the observations show that the energy can vary orders of magnitude over the scale of an eddy which makes parameterization of mixing by diffusion coefficients virtually impossible.

Another type of motion is the distinct, slightly elliptical rings often observed in the Gulf Stream area, previously called eddies,[21,22] but later suggested[23] to be called rings since distinct water types are involved in a circular configuration. The rings are formed when meanders of the Gulf Stream break away from the stream and form a ring of Gulf Stream water moving around a core of different water.[23,24] Rings can form both to the north and to the south. Those forming to the south or east of the Stream have a cyclonic rotation and carry cold water of slope origin into the Sargasso Sea. The cyclonic rings generally have an initial diameter of approximately 150 km which decreases with time, and surface current velocities about 150 cms^{-1}, barely decreasing with time. The central core reaches depths of about 2000m.[23] Such rings may form up to eight times a year, and normally exists for about 6 months but may last for a period of time up to 24 months. The main formation area seems to be between 60°W and 70°W[24] with speeds in the range 1 to 10 cms^{-1}.[23,25] Richardson et al.[26] used a drifting surface buoy tracked by satellite to follow a ring for several months. The ring moved northeast with an average speed of 9 cms^{-1} and eventually re-entered the Gulf Stream.

The anticyclonic rings are formed to the north or west of the Gulf Stream and enclose a warm core of Sargasso Sea water. They seem to interact strongly with the continental shelf and have lifetimes up to 12 months.[27] The diameter is about 100 km, surface current velocities are in the range of 30 to 75 cms^{-1} and the ring circulation is limited to the upper 1000 m.

When impacting on the shelf these rings generate a very effective exchange and renewal of water on the shelf.[27a,27b] In general the formation of rings effectuates a two-way transfer of water implying a large effective mixing.

The cyclonic rings serve as transporting agencies for water, substances, and marine communities into the Sargasso Sea.[28,29] In this capacity they also influence the distribution and transfer of pollutants. Although rings have by far been best documentated in the Gulf Stream area, rings of a similar nature have also been observed, among other areas, in the Kuroshio current system, at the northern boundary of the Antarctic Polar Front in Drake Passage,[30] and south of Australia. These are important features because they can transfer substances in substantial quantities from one water mass far into another water mass or area across zones which otherwise act as rather effective barriers against transfer. More studies are needed before the frequency of occurrence and the detailed characteristics of rings for other areas than the Gulf Stream area can be ascertained.

Over the continental rise and slope as well as over major topographic structures in the open ocean such as ridges, troughs, and seamounts the motion is influenced by topography. This is evident from meanders and eddies which may be related for instance to seamounts. At the continental border (rise, slope, and shelf) low-frequency fluctuations are often observed which may be due to combinations of various types of trapped waves.[6,14,31] More studies are, however, required to ascertain the types of waves involved and their precise characteristics. For a rigorous testing of the theoretical ideas long-term observations are required. Thompson and Luyten[32] analyzed the records from Site D, 39°N 70°W on the continental rise where motions with periods of 1 to 2 weeks and wavelengths of the order of 100 to 200 km have been found. They arrived at the conclusion that these waves were baroclinic, topographic Rossby planetary waves with energy decreasing upwards from the bottom. There was a marked minimum in the energy spectrum for periods in the range 8 to 4 days before the increase at the tidal periods, in agreement with theory.

Fluctuations of the motion with periods of the order of days, and maximum energy in the interval 1 to 8 days have been observed in the overflow from the Norwegian Sea to the Irminger Sea in the strait between Iceland and Greenland.[33] The oscillations appeared to be amplified in the downstream direction and were highly correlated in the southern part of the strait. Developing a quasi-geostrophic two-layer model for flow in a channel with sloping bottom, Smith,[33] using the observed values of relevant physical parameters, found instabilities of the model over a limited range of frequencies and wavelengths. The most unstable wave was 80 km long and had a period of 1 to 2 days, in good agreement with the observations. This type of instability is called baroclinic instability and appears to be a very important mechanism for converting potential energy into kinetic energy. Meso-scale fluctuations in the Mediterranean outflow have also been observed.[34]

Low-frequency oscillations in the current through the Florida Straits are well documented, with periods in the range from a few days to several weeks. Often these fluctuations are more energetic than the seasonal variations.[35] A mechanism which may cause these wave-like disturbances is the baroclinic instability,[36] but continental shelf waves may also play an important role.[37]

In concluding this presentation we can state that the observational efforts over the last few years have provided ample evidence for the general existence of meso-scale variability in the oceanic motion, with typical scales of the order of 100 km in space and days to weeks in time. This may be referred to as the oceanic weather. It is clear that this motion also influences the oceanic conditions to a considerable extent. Voorhis et al.[38] considered that large meridional and zonal bands on a scale of 40 to 400 km found in the sea surface temperature distribution in the North Atlantic subtropical convergence could be due to advective distortion by surface currents related to the deep baroclinic meso-scale eddy field. They found that wind-induced surface motion was not so important in generating the temperature features. They furthermore argued that jet-like shallow density currents could be important in advecting and distorting the surface temperature field on scales of 10 km and less. This being the case, the deep meso-scale motion will evidently also influence the distribution of pollutants not only in the deep water but also in the surface layer. The eddy motion is clearly important for the energy distribution and has also an influence on the large-scale oceanic circulation and hence the global transport of pollutants.

B. Inertial and Tidal Waves

However, there are obviously other types of motion in the ocean. Tidal and inertial motion generally occur in most areas. The inertial motions (Figure 5) are mainly wind-generated.[39] They constitute a very important part of the spectrum, at some distance from the coast, in areas of weak tidal motion, as in the Baltic Sea where they also were first discovered.[40] The inertial period $\pi/\omega_o \cdot \sin\phi$ depends upon latitude ϕ and the angular velocity of the rotation of the earth ω_o. Winds fluctuating with a period smaller than the inertial one are most efficient in generating this type of horizontal motion which starts at the surface and gradually penetrates into deeper layers without change of phase. The inertial oscillations also give rise to a large part of the small-scale horizontal shear in the ocean.

The tidal motion constitutes a very important energy source for mixing in the ocean, most efficiently on the shelf. In many shelf seas, as for example parts of the North Sea and the Irish Sea, the tidal mixing keeps the water column homogeneous throughout the year. The tidal motion generates elliptical paths which can degenerate into circular or linear paths, but the length of the excursion is generally small, of the order of kilometers. The motion is influenced by topography and stratification so that the direction along the elliptical path changes with depth. The tidal motion in shelf areas can often be very complicated with large variations over relatively small distances.

At the shelf break and the continental slope the oscillating tidal motion can generate internal waves in stably stratified water. These waves can propagate into the deep ocean and thus provide a mechanism for transfer of energy into the deep water. The

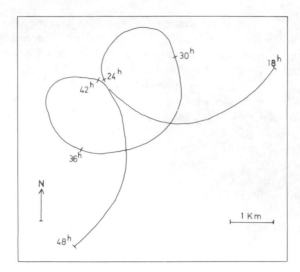

FIGURE 5.   Drift path of current cross at 8-m depth in
central Baltic Sea; observed inertial period 13^h54^m, theoretical period 14^h19^m. (After Gustafsson and Kullenberg.[40])

tides are therefore considered to be an important energy source also for mixing in the
interior of the ocean.[41]

Internal waves are very prominent features of the oceanic motion and manifest
themselves by considerable vertical oscillations of the density profile (Figure 6). The
waves are influenced by vertical variations of the stratification and by horizontal currents. Phillips[42] summarized recent observations and showed that the frequency and
wave number spectra had slopes near −2 over a fairly large range. The frequency ω is
limited to the range f ≤ ω ≤ N, where f is the Coriolis parameter and N the Brunt
Väisälä frequency. The amplitude can be considerable, up to several tens of meters.
Internal waves can be generated by slowly moving atmospheric pressure disturbances,
by an advecting pattern of surface stress, by flow over an irregular bottom topography,
and by an oscillating current (e.g., due to tides), encountering the continental
slope. On the basis of available observations, Garrett and Munk[43,44] suggested a universal internal wave spectrum for the deep water.

C. Small-Scale Motion

The occurrence of small-scale motion in the ocean, with time scales less than a few
hours, is much influenced by the density distribution. In stable conditions the vertical
motion is suppressed and the motion is essentially two-dimensional. However, three-
dimensional motion, which may be called true turbulent motion, can occur in the ocean
on a highly intermittent basis, both in time and space. This type of oceanic motion is
exceedingly difficult to observe and our information about it is therefore very limited.
The problem is partly to separate turbulent motion from internal wave motion.   A
few direct observations of the velocity field in the sea exist which have given information on the structure of the small-scale motion and on the occurrence of (three-dimensional) turbulence. Bowden[45] and Bowden and Howe[46] used an electromagnetic current
meter to study the velocity fluctuations close to the bottom (i.e., 50 to 175 cm above
the bed) as well as in the surface layer of tidal streams and estuaries. The data were
used to elucidate the distribution of kinetic energy. The average of the ratio $\overline{w'^2}/\overline{u'^2}$
varied in the range 0.17 to 0.26 and the average $\overline{v'^2}/\overline{u'^2}$ was 0.58 in the tidal stream.
Bowden found an average value of −0.38 for the ratio $\overline{u'w'}/(\overline{u'^2w'^2})^{1/2}$ and for the ratio

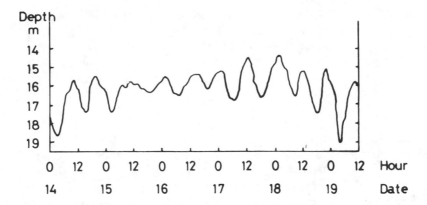

FIGURE 6.   Internal waves observed with a pycnocline follower in the central Kattegat. (After Kullenberg.[264])

E^2/w'^2, a value of 7.1. Here u', v' and w' are the fluctuating parts of the velocity components in the x, y, z directions respectively, and E^2 the fluctuating kinetic energy per unit mass. This implies that about 14% of the total fluctuating kinetic energy was contained in the vertical component. This agrees well with theoretical results given by Ellison[47] and Stewart.[48] Bowden and Howe[46] found that the Reynolds stress was considerably larger near the surface than close to the bottom and they considered that the influence of the bottom friction did not reach far into the water column. Later studies[49,50] in the ocean have also showed a thin (order of meters) bottom boundary layer.

Grant et al.[51,52] made a pioneering effort in measuring turbulence in and above the thermocline layer over the continental slope west of Vancouver Island using a hot film flowmeter. They found that in the surface layer at 15 m depth the motion was turbulent practically continuously while in and below the thermocline layer, at 90 m depth, the turbulent motion occurred in patches. These could have a horizontal extension of several tens or hundreds of meters while they were only of the order of 1 m thick. The turbulent motion was intermittent with a near-normal velocity distribution but with skewness due to the intermittency. The one-dimensional spectra of velocity and temperature fluctuations had an inertial subrange with a slope of $-5/3$ and fitted Batchelor's[53] spectrum in the viscous part with a slope of k^{-1}. The rate of energy dissipation per unit mass ε was found to be in the range $2.5 \cdot 10^{-2} - 5 \cdot 10^{-4} cm^2 \cdot s^{-3}$.

Nasmyth[54] continued these investigations in the waters off British Columbia towing sensors for temperature and velocity at depths from the thermocline to 330 m. He found that velocity fluctuations occurred in patches and were generally accompanied by temperature fluctuations. However, temperature microstructure was often observed outside the turbulent regions. Nasmyth suggested that this indicated that the region had been in turbulent motion not long ago; this feature is often referred to as fossil turbulence. The active patches of turbulence were layered, with layer thickness in the range of centimeters to meters and with sharp boundaries to more quiet and homogeneous regions. No clearly defined inertial subrange was found in the velocity spectra and they did not match the Kolmogorov spectrum too well. Nasmyth considered this to be an effect of the buoyancy forces. The rates of energy dissipation per unit mass were in the range $5 \cdot 10^{-3} - 2 \cdot 10^{-4} cm^2 \cdot s^{-3}$, conforming with other estimates.

Observations by moored current meters have shown that the current fluctuations normally have amplitudes that are larger than the time-averaged velocity. This implies that the Taylor hypothesis of frozen turbulence[55] cannot generally be applied, which makes it difficult to use Eulerian measurements to study the spectrum of the motion.

The interpretation of the turbulent motion beneath the surface layer of the sea as being characterized by a high degree of intermittency and with a patchy distribution seems to hold. This picture is also in agreement with the present understanding of the small-scale temperature and salinity distributions. An important implication is that these small- to micro-scale processes alone hardly will generate a vertical transfer in the interior of the ocean which is equivalent to a vertical mixing coefficient of 10^{-4} m^2s^{-1}.

D. Vertical Motion

The oceanic deep and bottom waters have low temperatures and the main sources of these waters are situated at high latitudes. The regions where sinking of dense surface waters occur are few and small; the main region appears to be in the Weddell Sea but there are also important regions in the North Atlantic, mainly in the Labrador Basin and the Norwegian Sea.[57] In these areas considerable vertical motion occurs during limited periods of the year transferring surface waters to the deep and bottom layers; i.e., penetrating up to thousands of meters. The newly formed deep water masses influence the deep water circulation in all the ocean basins, implying that from these regions substances can reach almost all parts of the ocean.

Part of the oceanic deep and bottom water is formed through overflows as in the Denmark Strait from the Norwegian Sea into the temperate Atlantic, and through flows generated at shelf regions flowing along the slope to the deep water, as in the Weddell Sea. During these flows a considerable entrainment occurs, implying an increase of the total volume so that a large amount of oceanic deep and bottom waters are formed. Another part of the deep and bottom waters are formed through deep convection as in the Labrador Sea.

In total, the deep and bottom water sources generate a volume of water which can force an upward motion in the interior of the ocean of about 10^{-5} cms^{-1} as an upper limit, assuming that the compensation is distributed evenly over the ocean interior. Many models of the oceanic heat balance, or of oceanic property distributions, have assumed that this upward motion continues in a horizontally uniform way up into the main thermocline. However, as has been shown by Needler[57a] this uniform motion does not necessarily occur.

Due to the difficulty of studying the high-latitude regions great attention has been given a third region, the Gulf of Lions in the western Mediterranean, where surface water sinks to about 2000 m depth during a limited period of some weeks in winter. These studies have shown that the bottom water there is produced during three phases.[58]

During the early part of the winter the density distribution takes a form of a dome in the Gulf coupled to a cyclonic (anticlockwise) circulation. The stability becomes much reduced in the center of the dome. This phase normally occurs during January.

During the second phase, which starts at the onset of the Mistral, violent vertical mixing takes place in a narrow region. The water column becomes well mixed down to about 2000 m over an area extending about 20 km in the N-S and 50 km in the E-W direction. Vertical velocities up to 10 cms^{-1} have been observed. During the third phase, starting with the end of the Mistral period, a fragmentation and interleafing of the water column occurs. At the surface eddies are formed with length scales about 15 to 20 km and speeds about 15 cm·s^{-1}. In general the water appears to drift slowly towards the south. This development shows that open sea convection can generate "rings" which, however, are of a different kind than the Gulf Stream rings.

In the open ocean deep, mixed layers can also form on a seasonal basis. Water from such deep mixed layers can spread along isopycnals and form subsurface layers like

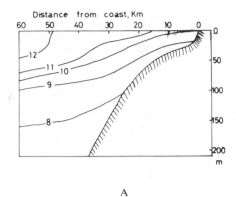

A

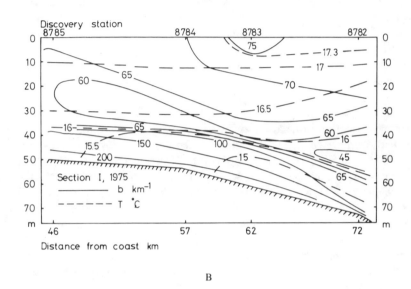

B

FIGURE 7. (A) Distribution of isothermes in a coastal upwelling section off Oregon. (After Smith.[265]) (B) Distribution of isothermes and light scattering in coastal upwelling section off northwest Africa.

intrusions in adjacent regions. An example is the North Atlantic central water mass.[58a,66]

An important type of upward motion is generated in certain boundary regions. Of these the major coastal upwelling areas along the west coasts of the continents are very important. The upwelling (Figure 7) is there generated by the combined action of the wind, the coast line, and the rotation of the earth. Due to the Ekman effect the surface water is deviated to the right (left) of the wind direction when looking downwind on the northern (southern) hemisphere. This implies that a divergence is generated near the coast for northerly winds in the northern hemisphere and southerly winds in the southern hemisphere. A shoreward compensating flow is generated at subsurface levels with upward vertical motion at the coast. The upwelling water comes from a few hundred meters depth, usually less than 300 m, implying that it is an important supply of nutrients to the euphotic zone. The upwelling areas are therefore generally characterized by a high primary production and rich fisheries.

Intensive research has been carried out in these areas during the last decade complemented with important theoretical studies. The observations have revealed the general features of the upwelling areas.[60] The isopycnal surfaces slope upwards towards the coast and reach the surface in a relatively narrow frontal zone separating cold, newly upwelled water from warmer water outside. Associated with the front there is strong equatorward flow about 50 m deep in the form of a coastal jet about 20 km wide. Centered around 200 m depth there is a poleward flow along the continental slope, about 300 m thick and about 50 km wide. The upwelling circulation is driven by the large-scale (of the order of 1000 km) wind systems, but shows large space and time variability and is very sensitive to changes of the wind direction and speed. The diurnal wind fluctuations due to the land-sea breeze also influences the upwelling. Variation in upwelling intensity is caused by topographic features such as the width of the shelf, capes, and submarine canyons through the shelf.

Vertical motion also occurs in areas of transient wind-generated up- or downwelling in coastal zones in semienclosed seas like the Baltic. In such stratified areas this type of motion can be especially important for the vertical exchange of pollutants and for the exchange between the coastal zone and the open sea.[61] Studies by Shaffer[62] have suggested that a large part of the total vertical flux across the permanent halocline in the Baltic Sea is due to coastal upwelling (or downwelling) generated by transient winds, especially during fall and winter.

E. Fronts

Important vertical motions are also associated with other frontal zones, both in shelf seas and in the open ocean. Of the latter the equatorial frontal system and the polar fronts can be considered the most important. These zones generally constitute areas of convergence with sinking of surface water. In the Southern Ocean the Antarctic Polar Front has been studied in some detail in limited areas.[63] At the southern part of the front, cool, relatively nutrient and particle-rich surface water sinks beneath the warmer northern water to depths between 800 and 1200 m forming the Antarctic Intermediate Water. In connection with the frontal zone considerable interleaving of cool and warm water masses occur, also evident from the distribution of suspended matter.[64] The layers (Figure 8) formed in this way can be from 10 to 100 m thick and can extend for considerable distances, at least several tens of kilometers. The isopycnals in this zone have a downward slope of about 10^{-3}, implying considerable vertical motion. The front extends around the whole Southern Ocean, implying that in this zone a considerable exchange of water and substances takes place between the surface layer and intermediate water layers. The generation of the front is related to the action of the wind and the rotation of the earth, but the precise nature of frontogenesis has not yet been clarified.

Also in the North Atlantic and the North Pacific similar frontal zones occur. The subarctic frontal zone in the Pacific Ocean around 40°N has been found to respond strongly to atmospheric forcing.[65] In this front horizontal gradients of temperature and salinity often balance each other so that the density gradients are weak, implying no pronounced baroclinic currents; the structure depends upon the season and is often highly complicated.[65]

Equatorial divergencies are generated by the combined action of the wind and the effect of the rotation of the earth which changes sign across the equator. A two-cell circulation (Figure 9) is generated with upwelling at the equator from about 100 m depth and downwelling around 4 to 5°N with upwelling again at about 10°N (in the Atlantic Ocean[66]). The circulation implies rich nutrient supply to the euphotic zone and a relatively high primary production in the equatorial zone. The convergence zone in the middle in combination with the equatorial current system also implies the possibility of accumulation of debris.

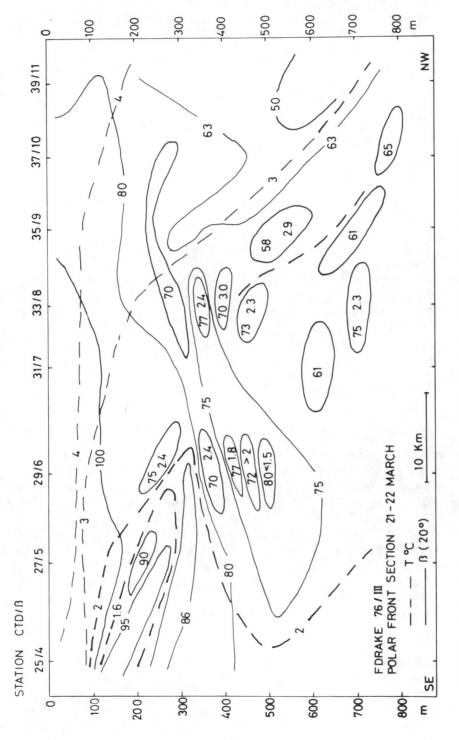

FIGURE 8.   Section showing temperature and light scattering distributions across the Antarctic polar front in Drake Passage.

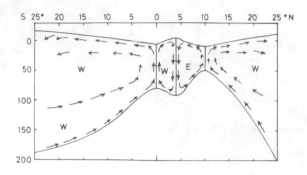

FIGURE 9.   Schematic representation of vertical circulation across Atlantic equatorial zone. (From Sverdrup, H. U., Johnson, M. W., Fleming, R. H., *The Oceans: Their Physics, Chemistry, and General Biology,* 1970, 635. Reprinted by permission of Prentice-Hall, Inc., Englewood Cliffs, N.J.)

Studies of frontal zones in open shelf seas have shown that they separate regions of strong vertical mixing from regions of weak vertical mixing.[68] It is clear that these fronts have an influence on the biological conditions and on the transfer of substances.[69] It is also clear that fronts are very common both in shelf seas and in the open ocean. Many of them are transient, occurring only during certain weather conditions or seasons whereas others are practically permanent, like the major oceanic fronts as well as several fronts in the northern European shelf seas and the Maltese front in the Mediterranean.[70]

III. FEATURES OF THE GENERAL CIRCULATION

A. Upper Layer

With the general circulation we understand the large scale, more or less permanent oceanic current systems driven by the combined action of wind and thermohaline forcing. The essentially wind-driven circulation of the upper layer has a similar appearance in all the oceans. There is a strong east-west asymmetry with the strongest currents at the western boundaries of the ocean. Such western boundary currents are the Gulf Stream in the North Atlantic, the Kuroshio in the North Pacific, the Agulhas Current in the South Indian Ocean and the East Australian Current in the South Pacific. These currents transport water poleward, with velocities in the range 100 to 200 cms^{-1} which has been transferred towards the west with the broad zonal flow in the north and south equatorial currents. Around 40° latitude the western boundary currents change to more zonal eastward flow. There is generally no pronounced currents at the eastern side, but rather slow equatorward flows feeding the equatorial currents. Thus so-called subtropical circulation gyres are formed[71] with an anticyclonic sense of rotation. The reason for the western asymmetry is to be found in the combined action of friction and the variation of the Coriolis (rotational) effect with latitude.[72] At higher latitudes there is a tendency towards a cyclonic gyre with intensified flow towards the equator along the western boundaries; examples are the East Greenland Current and the Falkland Current. The circulation reaches at least about 1000 m depth. At 800 m in the North Atlantic the circulation is very similar to the surface motion,[73,74] although the mean velocities are smaller, generally less than 30 cms^{-1}.

In the south all three oceans are connected to the Southern Ocean where the Antarctic Circumpolar Current flows around the Antarctic continent from west to east. This current reaches the bottom over much of its path and is strongly influenced by topography.

The wind-driven circulation in temperate and higher latitudes is essentially forced by the curl of the wind stress. The circulation is to a large extent modified by the rotation of the earth, the Coriolis force, and its variation with latitude. The necessary balance of the wind stress is obtained mainly by friction along the western boundaries, in the western boundary currents. The curl of the wind stress is also the forcing function for the vertical motion in the open ocean, over the top 1000 m. In the center of the anticyclonic gyres a downward motion along isopycnals is thus generated. This implies that the effect of the wind forcing reaches much greater depths than the direct frictional influence of the wind stress, which is limited to about 200 m.

A necessary forcing is also generated by the differential heating, which combines with the wind forcing. The downward motion generated by the wind stress curl can be balanced by a large-scale baroclinic circulation.

In the equatorial zone the surface circulation is generally westwards except for the Equatorial Countercurrent between roughly 4 and 10°N, slightly changing with season. An important feature is the Equatorial Undercurrent, a swift, narrow order of 100-km wide current flowing from west to east, in the Pacific Ocean at depths from about 300 m in the west to about 100 m in the east.[67] This current was first discovered in the Pacific but has since been found also in the other oceans. It is a major current with high persistency and transport. The mechanism of generation is not yet clearly understood but may be related to the pressure head generated by the surface transports from east to west.

Examples of transports in some major ocean currents are given in Table 1. Although the predominant currents are referred to as permanent, there are strong fluctuations, both seasonal and others. The meandering of the Gulf Stream and also the Kuroshio has been mentioned; the equatorial currents can even change direction. The picture described here is based on long-time averages, and short-time observations in any area may well give different results.

B. Deep Water Circulation

The mean deep water circulation is generally very different from the surface water circulation. An important feature is the undercurrent found at the western boundary in the Atlantic. The flow is southward under the Gulf Stream, and this flow has been found to extend across the equator to about 30 to 40°S. The current is due both to thermohaline forcing and effects of wind-driven circulation. The speeds are generally less than 20 cms^{-1}. In the North Atlantic there is also a southward flowing current along the eastern side of the mid-Atlantic ridge.

The circulation of the deep and bottom water in the interior is only fragmentarily known. Observations with neutrally buoyant floats show that swift, eddy-like motion can occur also at great depths. Attempts have been made to model the circulation numerically. Such models have suggested that the upward motion in the interior is not uniform. Other models, however, based on forcing by wind stress and thermohaline processes have often assumed a uniform upwelling motion in the interior determined by the amount of deep and bottom water formed at high latitudes. Veronis[75] found a uniform upwelling velocity of $1.5 \cdot 10^{-5}$ cm s^{-1}, in good agreement with other independent estimates.

Attempts have also been made to use observed three-dimensional distributions of various tracers such as salinity and oxygen.[76,77,78] The latter authors[78] applied the model of the thermohaline circulation given by Stommel and Arons[79] to the salinity and oxygen distributions in the Pacific Ocean. They found that an upwelling velocity of 3 m/year under the main thermocline, a horizontal eddy diffusion coefficient of $5 \cdot 10^2$ m^{2}s^{-1} and a vertical one of $6 \cdot 10^{-5}$ m^{2}s^{-1} gave a reasonably good fit to the observations. The tracer distribution was described by the mass balance equation

Table 1
WATER TRANSPORT IN SOME MAJOR
OCEAN CURRENTS

Current	Transport in $10^6 m^3 \cdot s^{-1}$	Main direction
Gulf Stream 40°N	75—115	NE
Kuroshio	40—50	NE
Florida Current	25—30	NE
East Australian Current	30—45	S
Canaries Current	15	S
Peru Current	20	N—NW
Pacific Equatorial Undercurrent	30—40	E
Southern Ocean Current		
Through Drake Passage	150	E
South of Australia	180	E

$$\frac{\partial C}{\partial t} = - \nabla \cdot (\underline{V} C - K_{ij} \nabla C) + J - \lambda C \qquad (1)$$

where C is the concentration, $\underline{V}$ the vector velocity, K_{ij} a diagonal matrix of eddy diffusion coefficients, J the regeneration flux, and λ a positive radioactive decay constant. The mixing was assumed to be isotropic in horizontal planes. With respect to the balance between advection, mixing, and regeneration Fiadeiro and Craig[78] concluded that

1.   The ratio K_h/K_z determines the depth to which the ocean feels the effects of the wind-driven circulation; a depth of 2.0 to 2.5 km requires a ratio of about 10^7.
2.   With a constant upwelling velocity a simultaneous increase or decrease of K_h and K_z changes the ratio of advection to mixing; a strong advection gives sharp gradients (on the normal to the flow) away from sources, strong mixing gives more linear gradients and more zonal distributions.
3.   The greatest residence time of the water is in the eastern equatorial region.
4.   Similar solutions may be obtained with different models.

Observed distributions of persistent anthropogenic substances such as PCBs have been related to known features of the circulation, as well as to some extent used to make inferences regarding the oceanic circulation.[80] These authors found relatively high concentrations of PCBs in the zonal belts between 40 and 55°N, between 10 and 20°N, and around 35°S in the Atlantic. In the South Atlantic the concentrations were generally very low. Although the evaporation properties of PCBs and the atmospheric transport pattern play a major role in explaining the observed distribution, the authors pointed out two main features of the North Atlantic circulation which also are relevant for explaining the distribution: (1) there is no net southward transport of surface water into the South Atlantic, and (2) the net northward flow of surface water is balanced by the southward deep water flow.[79] Low water temperature and the prevailing precipitation pattern would tend to keep the PCBs in the water column at high latitudes where the North Atlantic deep water is formed. The authors concluded that the net effect could be a transfer of PCBs across the equator to the South Atlantic and, eventually, to the Pacific Ocean.

The tritium introduced through atomic bomb tests has also been used to study transfer and mixing rates.[81,82] The latter authors found a southward migration of tritium introduced into the North Atlantic deep water (Figure 10).

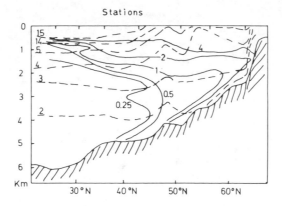

FIGURE 10.   Distribution of tritium (full drawn) and potential temperature (dashed, °C) in the North Atlantic from Denmark Strait to Sargasso Sea. (From Östlund, H. G., Dorsey, H. G., and Rooth, C. G., *Earth Planetary Sci. Lett.*, 23, 69, 1974. With permission.)

IV. SEMI-ENCLOSED SEAS

The current pattern in semi-enclosed seas and shelf seas is often more complicated than the pattern in the open ocean. The circulation in the mediterranean seas is generally driven by thermo-haline processes and by meteorological forcing including both the effects of wind stress and atmospheric pressure distribution. Marked seasonal variations often occur due to seasonal patterns in the meteorological conditions and the fresh water supply. The conditions in mediterranean seas depend critically on the water balance. A positive water balance as in the Baltic and Black Seas, will generate a stable stratification (Figure 11) with generally outflowing, brackish water in the surface layer and inflowing salty water in the bottom layer through the connections to the open ocean. Inside the constriction, vertical mixing can be weak and the renewal of the bottom water often occurs at intervals, although inflow of deep water is more or less continuous. The periodically stagnant conditions in the bottom water lead to oxygen depletion and often to anoxic conditions. Such regions can be sensitive to extra supply of organic material due to human activities. For such seas the residence time of the water is relatively long, e.g., about 35 years for the Baltic Sea, implying a potentially large build-up of pollutants.

A negative water balance, on the other hand, implies an increase of salinity in the surface water of the mediterranean sea relative to the outside ocean, due to the excess evaporation. Due to the increased density of the surface water, deep convection can occur in such areas during favorable meteorological conditions, normally a combination of cooling and winds. Thus the circulation will be reversed, with outflowing water in the subsurface layers and inflowing surface water through the connection to the open ocean. This is the case in the European Mediterranean Sea (Figure 12). In such areas the oxygen conditions in the deep and bottom waters are normally good. The circulation, however, also can lead to a concentration of substances floating on the surface such as oil, tar balls, and litter. This can also occur in fjords and estuaries with a negative water balance during part of the year.

The topography both of the constriction forming the connection to the open ocean and of the basin inside have considerable influence on the conditions. Especially the importance of shallow sills in preventing regular flushing of deep basins inside should be stressed. Selected semi-enclosed seas will be briefly discussed.

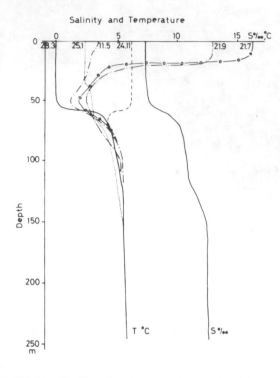

FIGURE 11.   Profiles of temperature in central Baltic Sea (57°20′N, 19°59′E) at different times of the year, showing development of seasonal thermocline, and one salinity profile for reference.

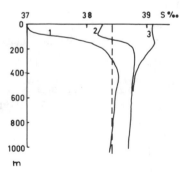

FIGURE 12.   Salinity profiles from west to east in Mediterranean Sea at 37°41′N, 5°57′E; 33°28′N, 19°59′E; and 34°13′N, 34°52′E. Dashed profile at 41°20′N, 4°49′E shows vertical homogeneity in northwestern Mediterranean Sea.

A. The Baltic Sea

In the Baltic Sea there is a weak cyclonal (anticlockwise) circulation in the surface layer with mean velocities of the order of 1 cms⁻¹. This flow is essentially thermohaline, due to the large annual fresh water supply to the Baltic Sea, being about 2.3% of the volume. The fluctuating part of the motion is dominating and primarily induced by the varying meteorological conditions, both winds and pressure gradients. The mean winds over the Baltic Sea are weak but the fall and winter storms are often both vigorous and persistent. The tidal currents in the Baltic are very weak, and the wind-induced inertial oscillations generally constitute the dominating part of the fluctuating motion.

The large river runoff to the Baltic generates a brackish surface layer with a thickness in the range 50 to 80 m. The salinity ranges from about 2 o/oo in the northern Baltic to about 7 o/oo in the southern parts. The incoming deep water, which has a tendency towards a cyclonic circulation around the basin, generates layers of more saline deep and bottom waters, with salinity in the range 10 to 13 o/oo in the central parts of the Baltic Sea. Occasionally strong inflows with some weeks duration can raise the bottom water salinity to 23 o/oo and 14 o/oo in the western and central parts, respectively. After such inflows, periods of up to about 5 years can pass until the bottom water is renewed. Generally the almost continuously inflowing deep water advects into the Baltic at its appropriate density level, often forming layers which can be traced over large parts of the area. A characteristic feature of the Baltic basin is the topographic division into several basins separated by sills or extended relatively shallow areas. The bottom water renewals therefore often occur in a series of pulses from basin to basin with a large part of the "new" bottom water being old water from the preceding basin.

The excess salt in the stagnant bottom water can only decrease by vertical mixing with the surface layer water. In the open parts of the Baltic the vertical mixing across the halocline layer is very weak[83] and therefore the role of the coastal boundary zone can be very important due to the more effective vertical transfer which can occur there. This was first pointed out by Walin[61] and has been confirmed by the measurements of Shaffer.[62]

B. Mediterranean Seas

In the European Mediterranean Sea the surface water flows in through the Strait of Gibraltar continuing eastwards along the southern coast but also with a northeastward branch into the Balearic Sea. The basic characteristic of the surface circulation is a series of cyclonic closed gyres, especially active in the northern parts of the Alboran Sea and in the Tyrrhenian Sea.[84] The Atlantic water is carried by these gyres into the Gulf of Lions. The remaining Atlantic water continues with the North African Current across the Straits of Sicily and passes into the Aegean Sea between Rhodes and the Turkish mainland. Because of the gyres a considerable amount of recirculation with limited exchange can be expected in this part of the Mediterranean. The limited information available from the eastern Mediterranean basin suggests a general cyclonic (anticlockwise) flow around the basin. Superimposed on this flow there appears to be at least two major gyres, one anticyclonic southeast of Crete and one cyclonic south of Greece and Turkey.[85] This implies a fairly closed circulation in the Levantine Sea which may also explain the strong thermohaline front which exists between the inflowing water of Atlantic origin and the existing older surface water. The Adriatic Sea, the only part of the Mediterranean with a positive water balance, shows a generally cyclonic surface circulation with southward motion along the Italian coast. However, also in this sea a number of more or less well-developed gyres occur which again implies a considerable recirculation.

The limited data available suggests that the summer and winter surface layer circulation is not basically different. The most important difference lies in the marked sinking motion which occurs in some areas, the Gulf of Lions and the Levantine Sea, during limited periods of the winter season.

Not much is known concerning the circulation in the intermediate layer between 250 and 400 m depth. Such a layer covers the entire eastern Levantine basin and a general westward motion is expected in the layer. Part of the water crosses the sills into the western basin and contributes to the intermediate water there.[86] The velocities are estimated to be low except in the Strait of Sicily where speeds up to 50 cms⁻¹ have been measured. Also in the deep waters a general westward motion is expected, although the sills of the Strait of Sicily effectively prevent high-density Levantine deep water from crossing into the western basin. In the Strait of Gibraltar there is a pulsating outflow of relatively dense deep water. This water sinks to its appropriate density level at about 1000 to 1200 m in the Atlantic and can be traced as a high-salinity layer far westward as well as northward.

Another important mediterranean sea is the Caribbean Sea situated in the trade wind belt. This area is also divided into a series of basins from east to west, and it is connected to the open Atlantic Ocean through a series of narrow passages. Most of these have small sill depths but some have sill depths in the range 1400 to 2200 m. In the top 1500 m the flow in the Caribbean is generally westward, with seasonal variations due to variations of the trade wind. The most well-developed current appears to be the zonal Caribbean Current flowing about 200 km north of the South American coast, turning northward in the western part and flowing out through the Yucatan Strait. Along the coast of Venezuela and Columbia the easterly winds generate coastal upwelling from about 200 m, leading to high primary production and important fishing. Below about 1500 m the circulation in the Caribbean is very weak due to the isolation of these water masses from the open Atlantic Ocean.

From the Caribbean Sea the water passes into the Gulf of Mexico. The Caribbean Current turns into the Yucatan Current which forms the western part of the large anticyclonic current, the loop current, occupying the eastern half of the Gulf. This current feeds the Florida Current into the Atlantic. The circulation in the western half of the Gulf is not as well defined, varying considerably in space and time. There are indications of a large, anticyclonic gyre, elongated in the northeast-southwest direction. It would lead too far to discuss separately the other mediterranean seas of which the most important ones are the intercontinental Arctic and Austral-Asiatic and the intracontinental Persian (Arabian) Gulf and the Red Sea.

V. FEATURES OF THE DISTRIBUTIONS OF PROPERTIES IN THE OCEAN

A. Vertical Distributions of Salinity and Temperature

The advanced instrumentation introduced into oceanographic work during the past decade has made it possible to study the structure of the vertical salinity and temperature distributions with a much better resolution than was previously possible. The layering, or small-scale structure, of the temperature and salinity fields occurring in many parts of the ocean was difficult to detect by classical oceanographic methods. Convincing evidence of the existence of these phenomena was brought about only by the introduction of *in situ* instruments recording simultaneously temperature, salinity (conductivity), and depth using small measuring heads, so-called STDs or CTDs. Several observations of oceanic microstructure (Figure 13) using such instruments have been reported since 1966.[87-89] These observations clearly show that the T-S fields are not smooth functions of depth but rather changing in a stepwise manner, with thin sheets

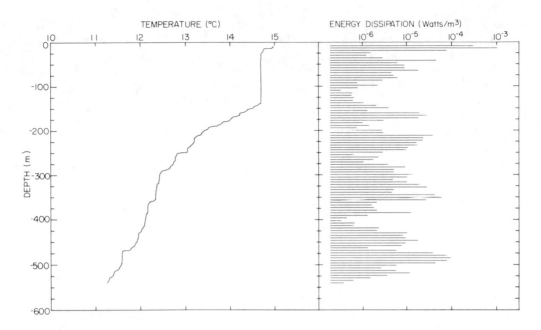

FIGURE 13. Temperature profile obtained with free-falling microstructure sensor in the North Atlantic south of the Azores. (From Osborn.[104]) The same instrument measures the velocity fluctuations making it possible to calculate the energy dissipation rate shown on the right-hand side, again from Osborn.[104]

of strong gradients separated by thicker layers of weak gradients. The thickness of the sheets is of the order 10 to 100 cm while the thickness of the layers in between is measured in meters. The horizontal extent of these features is measured in kilometers and the time scale has a magnitude of hours or days.

The early observations gave several suggestions as to possible generation mechanisms, such as sinking followed by advection along density surfaces with interleafing, and double diffusive effects, but it was concluded that more refined observations with a resolution of centimeters or less were required to solve the problem of the generation. Cox et al.[90] achieved a vertical resolution of about 1 cm by using a free-falling instrument. Measurements to about 1000 m depth in the Pacific Ocean showed the presence of series of temperature gradient spikes, about 10 cm thick and separated by a few meters. The temperature steps were proved to be sheets with a horizontal extent of at least 50 m.

Woods[56] and Woods and Wiley[91] have presented detailed measurements of the vertical structure of the seasonal thermocline in the central European Mediterranean, showing the existence of sheets of strong gradients 5 to 30 cm thick separated by layers of weak gradients of a few meters thick. Woods and Wiley also pointed out that the similarity between the microstructure observations in other parts of the ocean and those from the Mediterranean thermocline suggested similar generation mechanisms. Osborn and Cox[92] found very much the same temperature microstructure in such different parts of the ocean as the San Diego trough, the Florida Current, and the central Pacific Ocean. Howe and Tait[93] made detailed investigations in the Mediterranean outflow finding stepwise structure in the temperature and salinity distributions. Molcard and Williams[94] studied the deep-stepped (600 to 800 m) structure in the Tyrrhenian Sea and found 30 to 150-m thick homogeneous layers in a step-formed structure separated by about 6-m thick interfaces. The layers had great horizontal continuity with scales up to 100 km. Within the interfaces salt fingers (see below) were identified on scales of about 1 m, but salt fingers did not seem to be responsible for the step structure. Instead several other possibilities for the generation of the layers were suggested, such as mixing in the Strait of Sicily or as yet unknown mixing processes

Gregg[95] found temperature and salinity microstructure in the equatorial undercurrent around 35 to 65 m, i.e., above the velocity maximum. The structure of the profiles suggested ongoing strong vertical mixing, which is also in agreement with the general opinion that the undercurrent region is one of the relatively intense vertical mixing. Gregg found values of the Cox number (see below) above 3000 implying the presence of intense vertical turbulence. The Cox number is calculated from the model suggested by Osborn and Cox[92] of a balance between the production of temperature fluctuations by vertical turbulent motion and the smoothing of these fluctuations by molecular diffusion.

Introducing the vertical eddy coefficient for temperature K_{zT} and the rate of dissipation per unit mass ε_T

$$\overline{w'T'} = -K_{zT}\frac{d\overline{T}}{dz}, \quad \epsilon_T = -\overline{w'T'}\frac{d\overline{T}}{dz} = \kappa_T\overline{\left(\frac{dT'}{dz}\right)^2} \qquad (2a,b,c)$$

we obtain the relation

$$K_{zT} = \kappa_T\overline{\left(\frac{dT'}{dz}\right)^2} \Bigg/ \left(\frac{d\overline{T}}{dz}\right)^2 = \kappa_T \cdot Co \qquad (3)$$

where Co is the Cox number, κ_T the molecular diffusion coefficient for heat, w' the vertical fluctuating velocity, T' the temperature fluctuations, $\overline{T}$ the mean temperature, and z the vertical coordinate, positive upwards. The model leads to a value of K_{zT} but it should be noted that the assumed balance has not yet been independently proven. In the region of strong fluctuations, but with smooth profiles Gregg found a value of $K_{zT} > 5 \cdot 10^{-4}$ m²s⁻¹. In the main thermocline the K_{zT} values were around $5 \cdot 10^{-6}$ m²s⁻¹, whereas below the main thermocline at depths around 300 m, $K_{zT} \simeq 0.15 \cdot 10^{-4}$ m²s⁻¹.

Further observations of the oceanic small-scale distribution of salinity and temperature[96-98] have revealed that there is an increased microstructure intensity in the presence of horizontal intrusions. This would suggest that both horizontal and vertical processes are active in the formation of the structure. A recent summary of the basic features of oceanic microstructure has been given by Fedorov.[99]

Also in semi-enclosed areas and lakes a small-scale layered structure has been observed. Simpson and Woods[100] found that the thermocline in an open fresh water lake was divided in several isothermal layers about 5 m thick and separated by thin, sharp gradient layers. Fine structure was observed also below the thermocline where the mean gradient was very weak. A very convincing example of a layered structure has been observed by Hoare[101] in the Antarctic ice Lake Vanda.

Foldvik et al.[102] investigated the temperature field in a deep Norwegian fjord and found a structure very similar to that reported from other areas. Sheets of strong gradients, of the order of 10 cm thick, were separated by layers of weak gradients, or homogeneous, one or a few meters thick.

In the ocean, as well as in semi-enclosed seas, large quasi-horizontal layers of properties like heat, salt, suspended matter have also been found, which persist for very long time (months to years). Examples are outflows from the Mediterranean, inflows into the Baltic Sea, the Antarctic Intermediate Water formed at the Antarctic Polar Front. The existence of such layers, often having basin scales, suggests that the interior mixing can be very weak.

B. Vertical Velocity Distributions

Also the vertical distribution of the oceanic motion shows small-scale variability. This has been discovered by means of free-falling instruments measuring the horizontal velocity components from the surface to great depths.[103] Sanford[103] found variations in the current profile throughout the water column, with many layers of large time-variable shear superimposed on a smooth low-frequency profile. The vertical variations occurred over scales from about 10 to several hundred meters. The shear in the main thermocline region reached values of $5 \cdot 10^{-3}$ s^{-1} and in the deep water shears of about $1 \cdot 10^{-3}$ s^{-1} were found. Repeated profiles obtained at one location showed that the time-variability was mainly due to rotary motion of diurnal to inertial period. The coherence between profiles suggested a downward energy propagation, with a depth variation of the kinetic energy of the waves by and large in correspondence with the variation of the Brunt-Väisälä frequency. The Brunt-Väisälä frequency N is the frequency of oscillation of a small mass of fluid when displaced from its zero-order position in stably stratified conditions and then allowed to move freely. It is in the sea given by

$$N^2 = -\left(\frac{g}{\rho}\frac{d\bar{\rho}}{dz} + \frac{g^2}{c^2}\right)$$

where g is the acceleration of gravity, ϱ the density, and c the speed of sound. In many regions of the sea the latter term can be neglected. On the basis of velocity shears over 10-m intervals and a time mean Brunt-Väisälä profile Sanford found Richardson numbers in the range 0.5 to 4 over much of the water column. The Richardson number is defined as

$$Ri = N^2 \Big/ \left(\frac{du}{dz}\right)^2$$

where u(z) is the mean horizontal current velocity. Simultaneous profiles were also obtained at different horizontal distances, showing similarity for distances of about 100 m but becoming gradually more different for separations increasing up to about 10 km. Direct observations of vertical shears have recently been carried out by Osborn,[104] making it possible to calculate dissipation rates.

C. Vertical Distributions of Contaminants

It is quite clear that the small-scale distribution is a general feature of the oceanic salinity and temperature fields and not only in the main thermocline and seasonal thermocline layers but also in the deep waters. It is therefore not unreasonable to expect similar characteristics of the distributions of various contaminants. The presence of such features of the S, T distributions shows that the vertical mixing is generally very weak. Already in 1967, Folsom (cited by Okubo[73]) used an underwater nuclear test in the Pacific Ocean to show that the radioactive contaminant became distributed in thin layers, of the order of 1 to 10 m thick, at depths between 100 and 400 m.

Artificial dye tracers injected in and below thermoclines and haloclines have been observed to become distributed into sheets and layers of similar kind as those observed in the S, T distribution.[56,105,106] Kullenberg[106] carried out observations in the Baltic Sea area, in the North Sea, in the Mediterranean and in Lake Ontario as well as in various still fjords. The dye tracer was generally found to be distributed in sheets with thickness in the range 10 to 100 cm. These seemed to be connected to the temperature structure

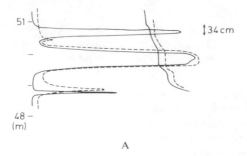

A

FIGURE 14.   (A) Profiles of dye and temperature (to the right) observed in the Baltic Sea after 12.7 (dashed) and 13.5 (full drawn) hours of dye tracing. (From Kullenberg, G., *Adv. Geophys.*, 18A, 339, 1974. With permission.) (B) Section through a dye patch in the Baltic Sea with profiles of dye and temperature at about 50 m, after 6 hr of dye tracing.

(Figure 14). In areas of weak stratification, as in Lake Ontario during the fall, thick layers with a stepwise structure were frequently observed.[107] The structure of these layers also seemed to be related to the temperature distribution, in the sense that a strong temperature gradient generally was accompanied by a sharp dye interface with no step structure. The step structure (Figure 15) was only found in very weak or neutral stratifications. The sheets observed in the thermocline layers were generally very persistent and could exist for several days after an injection of 10 kg of dye. The observations show that the small-scale structure is present in the water and that a contaminant will become distributed accordingly at its appropriate density level after a short time from the injection. This is quite clearly important for our understanding of the distribution and transfer of pollutants in the sea.

It should of course be noted that in well-mixed layers, such as in the wind-mixed surface boundary layer, the contaminants will also become mixed throughout the layer. This has been demonstrated by means of experiments where the dye was injected at the surface. In strong enough wind conditions the dye became mixed down to the bottom of the mixed surface layer.[5]

Radioactive isotopes such as Br 82 have also been used as artificial tracers in coastal waters.[108] Generally the observed distributions agree with the dye results, i.e., the contaminated water body becomes distributed in a structure which is clearly related to the salinity, temperature, and current structure.

The artificial tracers referred to behave as passive contaminants. Other contaminants may be active and may also influence the dynamic conditions, such as heat and salt released from different types of power plants, particles or a load of suspended matter released during certain mining operations or dumping into the sea, or fresh water discharge through sewage outfalls. Generally the direct influence on the motion is limited to an initial relatively short period during which the effluent adjusts itself to the distribution of density and motion in the recipient.[109]

Artifical tracers introduced into an initially small volume will at first be carried (advected) by the large-scale motion while the small-scale motion gradually acts so as to increase the size of the marked volume. At some stage, however, the "patch" will be torn apart by features such as the meso-scale eddies.

D. Suspended Matter

The particle distribution in the sea can be studied with a great advantage by means of light scattering techniques.[110,111] In several cases the particle content can be consid-

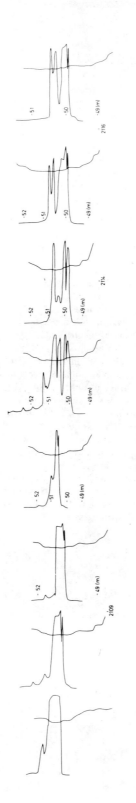

FIGURE 14B

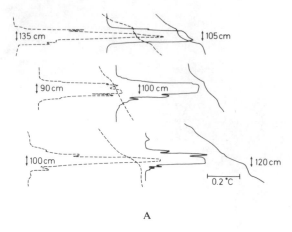

A

FIGURE 15.   (A) Profiles of dye and temperature observed in Lake Ontario at about 40 m (depth decreasing
upwards). Top left after 25 hr, top right after 39 hr of
tracing. Middle left after 22 hr, middle right after 38 hr
of tracing. Bottom left after 24.5 hr, bottom right after
40.5 hr of dye tracing. (From Kullenberg, G., *Adv.
Geophys.*, 18A, 339, 1974. With permission.) (B) Profiles of dye and temperature observed in Lake Ontario
between 20 and 25 m after 31.5 to 32 hours of dye tracing. (From Kullenberg, G., *Adv. Geophys.*, 18A, 339,
1974. With permission.) (C) Section through a dye patch
with profiles of dye and temperature from Lake Ontario
after 32.7 hr of tracing. (D) Hydrographic profiles of
T, S, σ_t, (top) and profiles of dye, injected in waste
water at about 25 m depth in Norwegian fjord, for different times after injection. (Adapted from Aure, J.,
Grahl-Nielsen, O., and Sundby, S., Spredning av olje-
holdig avløpsvann i Fennsfjorden fra olje raffinaderiet
pa Mougstad. Fisken og Havet, serie B, No. 2, Fiskeri-
direktoratets Havforskningsinstitutt, 1979.)

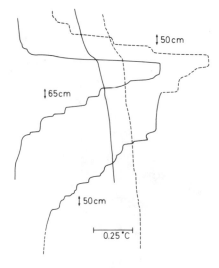

FIGURE 15B

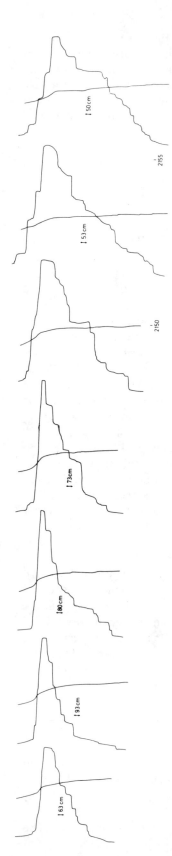

FIGURE 15C

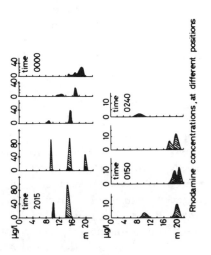

FIGURE 15D

ered as one characteristic of a given water mass and often there is a correspondence between particle content and salinity or temperature or both. Intruding water masses can often be traced also through particle distribution. Kullenberg[64] found that the cold subantarctic surface water south of the Antarctic Polar Front in the Drake Passage contained considerably more suspended matter than the warmer north of the Front (Figure 8). In the frontal zone intruding layers of cold water could also be identified on the basis of their relatively high particle content.

In coastal upwelling areas the particle distribution is influenced by the dynamics and to a certain extent also reflects the coupling between physical and biological processes. The upwelling water is poor in particles whereas the surface layer water becomes relatively rich in particles through high primary production. Kitchen et al.[112] demonstrated the importance of the mixing for the particle distribution also in an upwelling area.

It is quite common to find internal maxima of suspended matter. Those are often related to layers of relatively high stratification, since a particle maximum can be formed in layers where the mixing relative to layers above and below is reduced. Such maxima are for instance found in the Baltic halocline, in the Mediterranean intermediate water, in many fjords, in the subtropical underwater in the Gulf of Mexico, and have also been observed in the Arctic basin.

A large part of the suspended matter in the sea originates from land and the great rivers are very often important sources of particulate matter. The river plumes and the associate circulation can be observed by means of the particle distribution, as demonstrated by Jerlov[113] for the Po River plume in the Adriatic Sea and by Pak et al.[114] for the Columbia River plume in the Pacific Ocean. The latter study showed how the plume extended in a tongue-shaped form several latitude degrees southward from the source. Gibbs[115] showed that the turbid Amazon River water extends continuously in a particle-rich zone northwestward along the South American coast for about 2000 km; i.e., reaching to the Gulf of Paria. The width of the zone varied between 100 and 10 km, with a considerable band structure in the flow direction inside the turbid zone. In this case, as pointed out by Gibbs, the suspended matter should not be considered as a conservative property since sedimentation and other processes continuously influence the load of suspended matter. The study demonstrates that the suspended matter can be carried very far and presumably will influence the conditions also in the Caribbean. It is conceivable that the influence of the loads carried by the great Asian rivers can extend over similar distances.

In the deep and bottom waters variation of the particle content also occurs. Although much biological material becomes disintegrated in the top 500 m, some fraction will reach deeper layers. At least part of the atmospheric fall-out seems to reach the deep ocean floor without much change as suggested by the similar size distribution of quartz particles in atmospheric fall-out and deep sediments.[116] Particle-rich layers are often related to topographic features and in many areas of the ocean there is a near-bottom layer of relatively high particle content (Figure 16). Such layers are often referred to as nepheloid layers. Biscaye and Eittreim[117] found a layer thickness in the Atlantic of the order of 50 m on the Hatteras Abyssal Plain whereas the layer was of the order of 100 m thick in the regime of the western boundary undercurrent along the Blake-Bahama Outer Ridge. They also found that the characteristics of the bottom boundary layer water were interrelated. In the region of the undercurrent the current velocities were high, giving a high and varying particle load with values generally about $100 \ \mu g l^{-1}$. Over the abyssal plain the current velocities were much lower and the particle content less, of the order of $20 \ \mu g l^{-1}$ and more stable. In the water column immediately above the particle load was about $5 \ \mu g l^{-1}$.

Bottom layers with high particle content also occur on the shelf and slope. The upwelling area outside Northwest Africa was characterized by a well-defined bottom

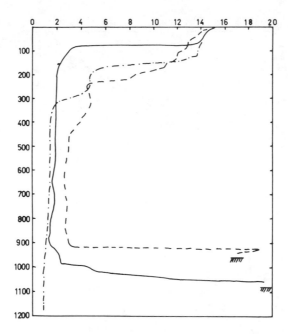

FIGURE 16. Profiles of light scattering (relative units) observed in the southern ocean. Full drawn at 48°41.3'S, 177°59.3'E, 1051 m; dashed at 49°16.5'S, 178°35.2'E, 942 m; dotted at 49°54.3'S, 179°17.9'E, 2975 m.

layer with a considerable amount of suspended matter relative to the water column above.[118] Over the shelf the layer was 10 to 20 m thick and over the slope 5 to 10 m thick. Such bottom layers of suspended matter can be generated by surface wave action, by strong currents, and by intense motions generated by internal wave action at the slope.

Also in semi-enclosed seas bottom layers of a high particle content can occur. In most parts of the Baltic Sea there is such a layer with a thickness of about 5 m. This can be due to the fairly high supply of material from both river runoff and primary production. The motion in the near-bottom layer is generally quite weak, only reaching eroding strength in areas of special topography and in some other areas during storm conditions. The importance of meteorological forcing in shelf areas should be emphasized and in particular strong wind conditions of some duration. Then both the directly wind-induced motion and the motion due to the wind-generated waves may reach the bottom and thereby generate very considerable sediment transport.

The suspended matter constitutes an important transport agency (see Volume II, Chapter 2). Hence it is important to realize that suspended matter can be transferred over large distances and that sediments can become resuspended.

Suspended matter introduced by rivers will tend to trap pollutants in the estuary. Flocculation occurs when the fresh water mixes with the seawater, implying gravitational settling of aggregates. These scavenge both particulate and dissolved material from the water column with at least a temporary deposition in the sediments as a consequence. Some particulate matter is transferred through the estuary to the open sea and resuspension also occurs in the estuary. However, reliable quantifications of these processes have not yet been achieved.

Mining, dredging, and dumping operations give rise to transfer of sediments and generation of suspended matter. The dispersion of the dumped material depends upon the properties of the material and the environmental conditions, i.e., stratification,

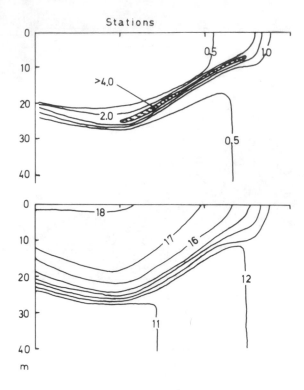

FIGURE 17.   Distributions of chlorophyll *a* in mgm⁻³ (top) and temperature °C (bottom) across seasonal front in shelf sea. (From Pingree, R. D., Holligan, R. M., and Mardell, G. T., *Deep Sea Res.*, 25, 1911, 1978. With permission.)

currents and wind. When dumping from a stationary vessel the initial dilution is of the order of 10^2. Part of the material becomes suspended in the water column and will disperse in layers as discussed above. Often the dumping is done from a moving ship which increases the initial dilution; for a speed of about 7 knots dilution values up to 10^3 can be obtained. Physical, chemical, and biological aspects of dumping are discussed in GESAMP.[119]

Considerable amounts of pollutants are introduced into coastal waters through various outlets. Experiments with tracers marking the injections have shown that the contaminants can become distributed in a layered structure. An example from a stratified fjord is shown in Figure 15D of a layer marked with dye and contaminated with hydrocarbons.

E. Horizontal Distributions

There are obviously also large horizontal variations in the distribution of oceanic properties. The salinity and temperature is influenced by exchange with the atmosphere, river runoff, and dispersion which result in very important large scale variations of a semipermanent nature.[120]

A characteristic feature of the S and T distributions is that often the variations are concentrated in frontal zones where the horizontal gradients thus are large relative to other parts of the ocean (Figure 17). This tendency is also found for other properties than S and T, such as nutrients and high biological productivity. This often leads to high fishing activities in frontal zones. The occurrence of fronts is coupled to the circulation and dynamical conditions. Often there is a tendency of convergence towards frontal zones which can lead to an accumulation of persistent substances in such areas.

For example, convergence zones between a river water plume and the more saline seawater often contain accumulations of floating material, including substances concentrated in the surface microlayer.

In recent years large research efforts have been directed towards the study of fronts, stimulated both by improved observational techniques such as the reliable very high resolution infrared radiometer developed for satellite remote sensing purposes, and by the realization that frontal zones are very common in the sea. Several different categories of fronts can be distinguished:[69]

1. Fronts of a planetary scale which may occur away from oceanic boundaries
2. Fronts representing the edge of the western boundary currents
3. Shelf-break fronts which may occur over the shelf break and part of the continental slope
4. Coastal upwelling fronts
5. River plume fronts in connection with major rivers discharged into the sea
6. Shelf sea fronts for instance formed in relation to estuaries or topographical and morphological features such as islands, capes, and shoals

The spatial extent of fronts can range from meters to several thousand kilometers and the persistence can vary from hours over seasons to years. Fronts can occur not only in the surface layer but also at depth and in the benthic (bottom) boundary layer.

VI. PROCESSES INFLUENCING THE S, T DISTRIBUTIONS

A. Vertical

The layered structure found in many areas beneath the region directly influenced by wind-induced mixing has a direct bearing on the dispersion conditions. In the interior of the ocean turbulence cannot be generated directly from the wind-influenced surface boundary layer although energy may radiate down through internal waves. The Richardson number based on the geostrophic motion is of the order of 10^4, but locally the Richardson number based on the scale of the layered structure may become less than the critical value.[121] Hence it is of interest to consider in some detail the generation mechanisms possibly responsible for the layered structure.

The remarkable similarity of the small-scale vertical structure of the salinity and temperature distributions observed in widely different regions suggests that similar generation mechanisms are active. In considering the generation processes it is pertinent to separate between the fine structure, order of meters (1 to 100 m) and the microstructure, order of centimeters (<100 cm). The fine structure can be generated through advective processes by interleaving or intrusion of water masses of different origin.[63,96,122] Supporting evidence for such a conclusion is the stability, persistence (order of days), and large horizontal extent (order of kilometers) of such layers. The interpretation of the fine structure as being due mainly to advective processes is based on observations primarily from coastal areas. However, the fine structure may also be generated by internal waves distorting for example a smooth temperature profile through vertical motion, as suggested by Garrett and Munk.[44] Observations by Hayes et al.[123] of temperature profiles in the northwest Atlantic give some support to the internal wave mechanism. Johnson et al.[124] made observations of the salinity and temperature in the open eastern Pacific Ocean on days of moderate and rough weather. In the first case the fine structure was transient and was suggested to be generated by internal wave distortions. The model spectrum of Garrett and Munk[44] was in good agreement with the observations. In the case of strong winds, in the same area, fine structure was also observed but the layers had fairly large horizontal extent (at least 6 km) and duration of at least several hours. Also in this case the authors suggest that the generation of the fine structure is dominated by internal waves but with an influence of vertical straining. The energy level was significantly higher on the day of strong winds than on the day of moderate winds.

In areas of large horizontal gradients of temperature or salinity, as in most frontal zones, the vertical internal wave motion may not be responsible for the fine structure.[125] Georgi[126] considered the possibility that low-frequency nearly horizontal internal waves could be responsible for fine structure observed in the Antarctic Polar Front zone. He found, however, that the horizontal displacements did not yield enough total variance nor the expected spectral shape, and therefore concluded that most of the observed fine structure may not originate from internal waves.

It can be concluded that both intrusion, or interleaving, and internal waves can generate fine structure, and that the type of mechanism dominating depends on the area and conditions. In connection with interleaving there is also a relatively large activity in the microstructure range,[95,98] which suggests a coupling. The microstructure is considered to be generated mainly by local processes.[122] Several mechanisms have been proposed which may be responsible for microstructure formation, and consequently also may be important for mixing in the sea. First, double diffusive convection can occur as a result of the different molecular diffusivities of salt and heat in areas of opposing salinity and temperature gradients. The large difference between the molecular coefficients gives rise to convective modes of motion transporting heat and salt although the density stratification is stable.[127,128] Double diffusion is a low Reynolds number process. In the case of a stable temperature and unstable salinity distribution, i.e., both properties decreasing with increasing depth, thin convection cells (salt fingers) occur in which descending elements loose their heat horizontally by diffusion more rapidly than their salt. Thereby they become heavier than their surroundings and continue to descend. The fingers can develop also in conditions of horizontal motion with vertical shear. Turner[129] has shown that the fingers occur in thin sheets separated by thicker layers which are kept homogeneous by vigorous convection driven by the salt transported through the fingers. Salt fingers have been observed in the ocean, on scales less than one meter, by means of an optical method[130] and by Magnell[131] using a towed instrument in the Mediterranean outflow.

Another type of double-diffusive convection occurs when the salinity distribution is stable and the temperature distribution is unstable, i.e., both temperature and salinity increase with depth. In this case buoyant elements use the potential energy of the temperature field to rise a finite height against the salinity gradient.[132] Layered structures of this type have been observed in the bottom waters of the Red Sea, in the Antarctic Lake Vanda, with salt water on the bottom and heating from the earth, and in the Arctic.[133]

Both types of motion have been discussed in detail by Turner.[134] The important point to note is that these processes will also transfer pollutants both when generated by the real salt and heat distributions and when generated by artificial changes in these distributions caused by human interference. For instance, this may occur when municipal sewage is trapped in a thermocline or halocline or when heated salty effluents are released. In principal the processes may occur in any component system of two or more where the components have differing molecular diffusivities.

A second mechanism possibly responsible for microstructure generation proposed by Phillips[42] and Munk[59] is the shear instability generated through the presence of internal waves. The maximum shear in internal wave motion occurs in the pycnocline layer. On the basis of this and a limiting critical Richardson number of ¼ for shear instability, Phillips,[42] found the limiting wave slope to be

$$(k\,a)_c = \frac{2f}{N_m} < 4$$

where k is the wave number, a the amplitude, N_m the maximum Brunt-Väisälä frequency, and f the Coriolis parameter. A local instability can develop when approaching the limiting slope, giving rise to a patch of intense, small-scale turbulence and mixing. The process is self-limiting since the mixing will reduce N_m and, hence, the occurrence of turbulent patches will be intermittent. Due to the stratification and the mean shear each patch will become horizontally elongated.

The mechanism appears very plausible in the light of several observations.[52,56,89,91] The flow visualization experiments by Woods[56] suggested that the breaking was due to short waves, of the order of 100 cm wavelength, triggered by longer waves, and that the vertical extent of the turbulent patch was in the range 3 to 30 cm. The internal wave shear possibly combines with the current shear to generate the shear instability. According to the criterium for shear or Kelvin-Helmholz instability given by Miles and Howard[135]

$$\lambda_c = 7.5 \cdot h$$

where λ_c is the critical wavelength and h the thickness of the shear zone, which Woods[56] found to be in fair agreement with his observations. The breaking waves generate so-called billows. On the basis of a maximum height of 30 cm for the billows, Woods[136] estimated that the Reynolds number in the billows is of the order of 10^2 to 10^3. The billows occurred intermittently and not more than 10% of the flow in the thermocline layer was turbulent at any time.[91]

A third mechanism was proposed by Orlanski and Bryan[137] and is related to the shear instability mechanism in that internal waves are the primary originators of the turbulence. The instability is however gravitational and occurs when the particle velocity of the wave motion exceeds the phase velocity of the wave. This can take place before the shear instability occurs, also when rotational effects are included.[138]

Several laboratory studies have been carried out of the various types of instabilities which can occur in a field of internal waves. Recently Thorpe[139,140] has shown that the presence of shear in the basic flow may considerably reduce the slope at which breaking can occur, and that the Reynolds stresses very efficiently transfer wave energy into mean motion. This implies that only a fraction of the initial wave energy is available for transfer to potential energy when breaking occurs. Thorpe[140] also suggested that internal wave energy may become concentrated in regions of maximum velocity through the action of low frequency variability, for example the mesoscale eddies.

B. Horizontal

Fronts occur remarkably frequently in the ocean. Recently the improved satellite remote sensing technique in the infrared spectral range with very high resolution has been used to identify temperature fronts in most areas of the oceans.[141] Using such data in combination with wind data Roden and Paskausky[142] showed that the influence of the wind through the Ekman effect could explain at least part of the differential advection (on time sacles up to a weak) required for frontogenesis in the surface layer of the open north Pacific Ocean.

Frontal zones are frequently coupled to topographic features. The frontal boundary formed between the warm and saline Atlantic water flowing into the Barents Sea and the outflowing cold and less saline Arctic water covers a distance of about 1500 km. The front makes a sharp bend around the south of the Bear Island, and it was suggested by Johannesen and Foster[143] that the position of the front in the Barents Sea is locked to the outer part of the shelf, in general following the 100-m isobath. Fronts often occur over the shelf break and slope, and a possible reason for this is the change of mixing conditions when passing this topographic boundary. Over the shelf, motions due to tides and winds can generate a relatively strong mixing which is considerably weakened in the deeper water outside the slope. The transition from the broad shelf in the eastern Bering Sea is marked by a nearly 1000 km long, diffuse and persistent (order of years) haline front over the continental slope.[144] They considered the front to be due to the decrease of the tidal mixing efficiency with the rather sharply increasing depth over the shelf break and slope. In the near coastal zone the horizontal eddy diffusion coefficient for salt K_h was estimated at $3 \cdot 10^2$ m^{2}s^{-1}. Kinder and Coachman[144] remarked that in general similar fronts can be expected in areas with a broad shelf, significant river runoff, motion on the shelf dominated by high frequencies (relative to seasonal changes) and the absence of a strong boundary current. Another front of this type could be the front between the shelf water and the slope water south of New England.[145]

The motion and mixing conditions in the New England front during the summer period were studied by Voorhis et al.[146] They found mean westward motion along the shelf between the 80- and 300-m isobath, completely confined to the shelf water. They could not detect any mean cross-shelf or vertical motion. Considerable interleafing, with layers about 10 m thick, was observed combined with strong microstructure. The interleafing layers were thinned by the vertical shear loosing both heat and salt on a time scale of 1 to 3 days. The processes suggest efficient vertical exchange resulting in a transport of salt from the slope water to the shelf water. The authors estimated the vertical eddy diffusion coefficient for salt K_{zs} through the relation $K_{zs} = h^2/t$ to $5 \cdot 10^{-4}$ m$^2 \cdot$s^{-1} with the thickness h $\sim$ 10 m and the time scale t $\sim$ 2 days. From this the vertical salt flux was found and the total input of salt to the shelf water, assuming a 10-km wide front. The amount turned out to be a significant part of the annual salt budget of the shelf water mass.

In winter this strong, narrow front separates cold, fresh shelf water from warm, salt, and denser water over the slope. The front shows wave like distortions with wavelengths of about 80 km along the shelf. Flagg and Beardsley[147] showed that the bottom slope had a strong influence on the stability of the front, in general tending to increase the stability. This implies that some mechanism other than local baroclinic instability must be responsible for the generation of the frontal undulations. On flat topography, however, the front would be quite unstable and Flagg and Beardsley[147] suggested this could be part of the explanation why such persistent fronts are situated near the shelf break. This mechanism would act in the same sense as the mixing mechanism discussed by Kinder and Coachman.[144]

This type of fronts are marked boundaries for water properties including nutrients, plankton and biological activity in general, as well as concentration of pollutants, and they influence the exchange between coastal waters and the open sea. They may also act as traps of pollutants.

During the warm season fronts can also occur on the shelf, marking the boundary between stratified and vertically mixed regions; such fronts have in particular been studied around the United Kingdom. Simpson and Hunter[148] showed that fronts could be generated by variations in the level of tidal mixing and that the locations of the fronts were essentially determined by the parameter H/U^3, H being the water depth and U the amplitude of the tidal current velocity. For large values of H/U^3 the tidal mixing is not able to fully mix the water column. Thus there is a critical value of H/

U^3 which was found to be in the range 50 to 100 m²s⁻³ for the shelf seas in question. In a later model Simpson et al.[149] included both wind mixing and tidal mixing, assuming a local energy balance. The potential energy per unit area, P_E, relative to the mixed state

$$P_E = \int_{-H}^{o} (\rho - <\rho>) \, g \, z \, dz \, , \quad <\rho> = \frac{1}{H} \int_{-H}^{o} \rho \, dz \qquad (4)$$

was used as an index of stratification. Fixed fractions of the boundary stresses due to the tidal motion and the wind were assumed to be driving the mixing. The available energy per unit time, P_A, for mixing was given as

$$P_A = c_b \, k_b \, \rho \, \overline{U_b^3} + c_d \, k_S \, \rho_S \, \overline{W_S^3} \qquad (5)$$

where k_b and k_s are mixing efficiencies, c_b and c_d are drag coefficients, W_s the wind speed near the surface, and U_b the bottom velocity. Surface heating will tend to increase the stratification at a rate

$$\left(\frac{dP_E}{dt}\right)_{heat} = -\frac{\alpha g \cdot \dot{Q} H}{2 c_p} \qquad (6)$$

where $\dot{Q}$ is the rate of heat input, c_p the specific heat, and α the volume expansion coefficient. The local potential energy balance then becomes

$$\left(\frac{dP_E}{dt}\right) = \frac{-\alpha g \cdot \dot{Q} H}{2 c_p} + c_b \, k_b \, \rho \, \overline{U_b^3} + c_d \, k_S \, \rho_S \, \overline{W_S^3} \qquad (7)$$

The ratio R will then determine the position of the front

$$R = -\frac{dP_E}{dt} \bigg/ \frac{\alpha g \cdot \dot{Q} H}{2 c_p} \qquad (8)$$

the critical value being R = O. The ratio R represents the fraction of the potential energy input by heating which is retained in the water column. The critical condition specifying the position of the frontal zone is R = O, and for R > O the stratification will increase with time. The case R < O is not included in the model since P_E > O corresponds to an unstable stratification. Simpson et al.[149] found that both mixing sources (tides and winds) were important and that the model gave a satisfactory qualitative account of the distribution of stratification and occurrence of frontal zones on the shelf. The detailed structure of the front could not be predicted. Observations showed the presence of relatively large-scale (∼25 km) instabilities. Further studies of the fronts in the European shelf seas have been carried out by Pingree[68] and Pingree and Griffiths.[150] These fronts act as important seasonal boundaries between areas of high and low primary production. They can also act as traps for pollutants. In conclusion, fronts in general should be carefully considered in pollution assessment studies. A preliminary overview of the occurrence of fronts in various areas can often be obtained by means of remote sensing techniques.

VII. GENERAL EQUATIONS, PARAMETERS, AND SCALING

A. General Equations

The general equations governing the problem of transfer and transport in the sea have been discussed in many recent works.[42,151] For large-scale circulation studies a spherical coordinate system should be used but for the consideration of relatively small- or meso-scale processes a Cartesian coordinate system is used with x-, y- and z-axes positive eastwards, northwards, and upwards, respectively. The Boussinesq approximation is normally used; i.e., (1) the density is replaced by a constant value when multiplied with the acceleration terms, (2) the continuity equation is replaced by the condition of incompressibility, and (3) the density is assumed to be a linear function of salinity and temperature. The last condition is only valid over a not too large range of salinity and temperature. The equations are

$$\frac{\partial u}{\partial t} + u\frac{\partial u}{\partial x} + v\frac{\partial u}{\partial y} + w\frac{\partial u}{\partial z} - fv$$
$$= -\frac{1}{\rho_o}\frac{\partial p}{\partial x} + \frac{\mu}{\rho_o}\left(\frac{\partial^2 u}{\partial x^2} + \frac{\partial^2 u}{\partial y^2} + \frac{\partial^2 u}{\partial z^2}\right) \tag{9}$$

$$\frac{\partial v}{\partial t} + u\frac{\partial v}{\partial x} + v\frac{\partial v}{\partial y} + w\frac{\partial v}{\partial z} + fu$$
$$= -\frac{1}{\rho_o}\frac{\partial p}{\partial y} + \frac{\mu}{\rho_o}\left(\frac{\partial^2 v}{\partial x^2} + \frac{\partial^2 v}{\partial y^2} + \frac{\partial^2 v}{\partial z^2}\right) \tag{10}$$

$$\frac{\partial w}{\partial t} + u\frac{\partial w}{\partial x} + v\frac{\partial w}{\partial y} + w\frac{\partial w}{\partial z}$$
$$= -\frac{1}{\rho_o}\frac{\partial p}{\partial z} - g\frac{\rho}{\rho_o} + \frac{\mu}{\rho_o}\left(\frac{\partial^2 w}{\partial x^2} + \frac{\partial^2 w}{\partial y^2} + \frac{\partial^2 w}{\partial z^2}\right) \tag{11}$$

$$\frac{\partial u}{\partial x} + \frac{\partial v}{\partial y} + \frac{\partial w}{\partial z} = 0 \tag{12}$$

$$\frac{d\rho}{dt} = 0 \; ; \; \rho = \rho_r\left[1 - \alpha(T - T_r) + \beta_s(S - S_r)\right] \tag{13}$$

$$\frac{\partial T}{\partial t} + u\frac{\partial T}{\partial x} + v\frac{\partial T}{\partial y} + w\frac{\partial T}{\partial z} = \kappa_T\left(\frac{\partial^2 T}{\partial x^2} + \frac{\partial^2 T}{\partial y^2} + \frac{\partial^2 T}{\partial z^2}\right) \tag{14}$$

$$\frac{\partial S}{\partial t} + u\frac{\partial S}{\partial y} + v\frac{\partial S}{\partial y} + w\frac{\partial S}{\partial z} = \kappa_S\left(\frac{\partial^2 S}{\partial x^2} + \frac{\partial^2 S}{\partial y^2} + \frac{\partial^2 S}{\partial z^2}\right) \tag{15}$$

where u, v, and w are the velocities in the x, y, and z directions, respectively; T is temperature; S is salinity; ϱ is density; p is pressure; and t is time. The molecular transport coefficients of momentum, heat and salt, are given by μ, κ_T and κ_S, respectively, and ϱ_r, T_r, S_r denote a reference state defined as an ocean with constant potential temperature and density which is in rest relative to the rotating earth. In the upper ocean the reference density differs only slightly from the density ϱ_o at the surface, so in practice for the surface layer $\varrho_r \equiv \varrho_o$. The vertical pressure gradient defined by the reference state, i.e.,

$$\frac{\partial p_r}{\partial z} + g\rho_r = 0 \tag{16}$$

can be subtracted from the actual gradient. Thus the buoyancy term b_* is introduced,[19] often also called the reduced gravity g′.

$$b_* = \frac{-g(\rho - \rho_r)}{\rho_r} = \frac{g\,\Delta\rho}{\rho_r} \tag{17}$$

It is costumary to introduce the Reynolds assumptions that the respective fields can be expressed as the sum of a mean and a fluctuating component,

$$u, v, w = U, V, W + u', v', w'$$

$$p = \bar{P} + p', \quad \rho = \bar{\rho} + \rho'$$

where

$$\overline{u'} = \overline{v'} = \overline{w'} = \overline{p'} = \overline{\rho'} = 0$$

The mean motion is interpreted as the time average of the motion over a number of sequences typical of the relevant turbulence or fluctuations; i.e., over a time long compared to the typical time scale of the dominating fluctuating motion.

Introducing Reynolds assumptions into the equations and averaging, results in additional terms due to the correlation between the fluctuating components. The so-called Reynolds stresses

$$-\rho\,\overline{u'_i u'_j}$$

occur in the momentum equations; u'_i and u'_j are the fluctuating velocity components in the i and j directions, respectively, and there are nine different correlations of this type.

The correlation terms are unknown and their effect can either be taken into account by means of eddy coefficients or by formulating a more elaborate turbulence model. The latter requires assumptions regarding structure and characteristics of the turbulence.[1,2]

The energy equations for the fluctuating motion are derived by multiplying Equations 9, 10, and 11 with the respective component of the fluctuating motion, and averaging. Assuming horizontal homogeneity of the fluctuating field the equations become

$$-\frac{\partial}{\partial t}\frac{1}{2}\overline{u'^2} = \frac{\overline{u'}}{\rho_o}\frac{\partial p'}{\partial x} + \left[\overline{u'^2}\frac{\partial U}{\partial x} + \overline{u'v'}\frac{\partial U}{\partial y} + \overline{u'v'}\frac{\partial U}{\partial z}\right] \tag{18}$$

$$+ \left[\frac{\partial}{\partial t}\frac{1}{2}\overline{w'u'^2} - \frac{1}{2}\overline{u'^2\frac{\partial w'}{\partial z}}\right] + 0 - \frac{\mu}{\rho_o}\overline{u'\,\nabla^2 u'}$$

$$-\frac{\partial}{\partial t}\frac{1}{2}\overline{v'^2} = \frac{\overline{v'}}{\rho_o}\frac{\partial p'}{\partial y} + \left[\overline{u'v'}\frac{\partial V}{\partial x} + \overline{v'^2}\frac{\partial V}{\partial y} + \overline{v'w'}\frac{\partial V}{\partial z}\right]$$

$$+ \left[\frac{\partial}{\partial z} \frac{1}{2} \overline{w'v'^2} - \frac{1}{2} \overline{v'^2 \frac{\partial w'}{\partial z}} \right] + 0 - \frac{\mu}{\rho_0} \overline{v' \nabla^2 v'} \qquad (19)$$

$$- \frac{\partial}{\partial t} \frac{1}{2} \overline{w'^2} = \overline{\frac{w'}{\rho_0} \cdot \frac{\partial p'}{\partial z}} + 0$$

$$+ \left[\frac{\partial}{\partial z} \frac{1}{2} \overline{w'w'^2} - \frac{1}{2} \overline{w'^2 \frac{\partial w'}{\partial z}} \right] \qquad (20)$$

$$+ \frac{g}{\rho_0} \overline{w'\rho'} - \frac{\mu}{\rho_0} \overline{w' \nabla^2 w'}$$

The first terms on the right hand side express the redistribution of turbulent energy between the velocity components through the action of the pressure fluctuations. The second terms are the production or Reynolds stress terms, representing the interactions between the fluctuating and the mean fields. This term is absent in the vertical component equation. The third terms give the vertical turbulent diffusion of the quantity. The corresponding horizontal components are absent due to the assumption of horizontal homogeneity. The effect of this term is a redistribution in space of the quantity. The fourth terms represent the effect of vertical stratification and is only present in the vertical component equation. Finally, the fifth terms represent the molecular effects.

In applications several terms in these equations often have to be disregarded simply because of lack of information about them. Monin and Yaglom[151] suggest that for a parallel shear flow in the absence of a destabilizing heat flux, the pressure terms and the diffusion terms may be disregarded. In the steady state, with vertical shear in the x direction, the total energy balance for the fluctuating motion is then given by

$$\overline{u'w'} \frac{dU}{dz} + \frac{g}{\rho_0} \overline{w'\rho'} + \epsilon = 0 \qquad (21)$$

B. Quantification of the Effect of Fluctuations

The eddy viscosity (eddy coefficient) models have given satisfactory predictions for many two-dimensional thin shear flows. However, for three-dimensional flows and other flows with more than one component of mean velocity gradient the approach has not been satisfactory. For solving the equations governing three-dimensional flows powerful numerical techniques have been developed. This has in turn led to considerable advance in the understanding of turbulence models. One class of turbulence models is based on the solution of the approximated equations for the kinematic Reynolds stress $-\overline{u'_i u'_j}$ (see, for example, Launder et al.[152]). The equations governing the transport of the Reynolds stresses will contain correlations between velocities and velocities and pressure and velocity triple correlations.

The aim is to convert these equations together with the equations for the mean motion and the continuity equation into a closed set of equations for mean velocities and Reynolds stresses. The problem is to represent the turbulence quantities as empirical functions of the mean velocities, the Reynolds stresses, and their derivatives. It is usually assumed that the Reynolds number is large, and it is necessary to consider separately free flows away from boundaries (walls) and flows close to walls.

A main part of the problem concerns the approximation of the pressure-strain correlation. The pressure fluctuations give rise to a tendency towards isotropy in the turbulent motion. The difference between the homogeneous free shear layer and the near-wall flow lies in the relative magnitudes of the Reynolds stresses; near the wall the streamwise component is larger and the transverse component is smaller than for the free layer. This implies that slightly different approximations have to be used.[152]

The set of equations obtained through the approximations is used to give predictions of various characteristics of the flow which can be compared with direct observations, thus testing the "goodness" of the model approximations. There are clearly a number of difficulties both in regard to the theoretical approximations and in regard to the measurements. The measurements are often extremely difficult. It seems safe to conclude that as yet only carefully executed laboratory experiments can be used for testing the turbulence closure models and that the conditions in the sea, and our knowledge about them, are such that as yet it is very difficult to use the results in oceanic applications.

Additional complications are encountered when buoyancy or gravitational effects must be included. Again second-order closures can be used considering also the transport equations for instance for the heat flux correlations $\overline{u'_iT'}$ (e.g., Launder[153]).

Also in this problem the pressure-containing correlations are important. Launder[153] generalized the model of Launder et al.[152] to include buoyancy effects. In particular he stated that the direct influence of gravity on the pressure-containing correlations should be included in the calculations.

This implies that one effect of the pressure fluctuations is to make the effective stress- and heat-flux generation tensors more isotropic. Under the assumption of vertical variation only of temperature and horizontal mean velocity, Launder found that the effect of buoyancy on the turbulent fluxes imply a gain in the level of streamwise fluctuations at the expense of the vertical fluctuations. The relative magnitude of the lateral fluctuations was unaffected. The changes in intensity depended upon the ratio $Rf/(1 - Rf)$, where Rf is the flux Richardson number, i.e., the rate at which turbulent energy is removed by working against the gravitational field divided by the rate at which it is generated by the mean shear (see also below). Launder presented his theoretical results against the Richardson number Ri and found a fair correspondence with observations by Webster.[154]

The influence of a horizontal boundary (wall) was considered by Gibson and Launder[155] with particular reference to the atmospheric boundary layer. The influence of the wall on the pressure-strain correlations is normally related to the ratio ℓ/z of the turbulence length scale to distance from the wall. As long as ℓ increases with distance, the wall effect is constant and where ℓ levels off in the outer part of the shear flow the wall effect decreases. In the atmospheric boundary layer the ratio ℓ/z has been found to be very sensitive to the strength of the stratification (see review by Turner[134]). With increasing Richardson number the ratio ℓ/z falls markedly. Recently it was discovered that also in turbulent motion ordered structures can be present in the form of coherent three-dimensional patterns referred to as bursts (Kim et al.[156]), and well-defined large structure rollers (Brown and Roshko[157]).

The effect of the fluctuating motion which we as yet cannot describe in a satisfactory way is in applications normally parameterized by using the Reynolds analogy and define turbulent or eddy transport coefficients for momentum A and for heat, salt, or mass K. The reason for the use of such apparent turbulent transport coefficients is that they make it possible by hypotheses to express the transports due to the fluctuating, relatively small-scale motion in terms of parameters which may be determined from mean, relatively large-scale conditions. Since the horizontal and vertical fields of motion are basically different in the sea the coefficients representing horizontal and vertical transports are also different. The coefficients on the main axis are defined by

$$A_x \frac{\partial V}{\partial x} = -\overline{u'v'}, \ A_y \frac{\partial U}{\partial y} = -\overline{u'v'}, \ A_z \frac{\partial U}{\partial z} = -\overline{u'w'}$$

$$\text{(22)}$$

$$K_{zT} \cdot \frac{\partial T}{\partial z} = -\overline{w'T'}, \ K_{xT} \frac{\partial T}{\partial x} = -\overline{u'T'}, \ K_{yT} \frac{\partial T}{\partial y} = -\overline{v'T'}$$

The analogy to the molecular theory is evident (see Okubo[73] and this volume Chapter 2).

C. Scaling Approaches and Parameters

One of the basic objections to the use of turbulent transport coefficients is that it allows only for transports down a gradient. This may be satisfactory as long as the transports are produced by fluctuating motion, often called eddies, with a characteristic length scale small compared to a distance over which the mean gradient changes appreciably. The transport mechanism should have a characteristic length much smaller than that of the mean field. In the mixing length theories the coefficients are determined by the root mean square turbulent velocity and a length scale characteristic of the turbulent large-scale structure.[158] In several applications, however, the existence of such coefficients is questionable because no finite characteristic length scale can be defined. For global conditions the flux is carried by relatively large-scale perturbations and then the transport can be countergradient. This has been found to be the case in the atmosphere and recent studies seem to confirm this to be the case also in the ocean as regards momentum and energy transport.[7]

In large parts of the atmosphere and the ocean the flow is approximately two dimensional. It is also nearly in a state of hydrostatic and geostrophic balance, the latter implying an approximate balance between Coriolis and pressure gradient forces in the horizontal. The fluctuating part of such motion is sometimes called geostrophic turbulence. The characteristic properties are very similar to purely two-dimensional turbulent motion. For such flows the mean-squared vorticity (or twice the enstrophy) is conserved. This implies that transfer of energy from one wave number to higher wave numbers is accompanied by a larger transfer of energy to lower wave numbers.[159] On the basis of this, Kraichnan[160] postulated that an inertial subrange exists for two-dimensional turbulence, where energy injected in a given wave number band will be transferred uniformly to lower wave numbers while vorticity is transferred uniformly to higher wave numbers. This leads to a $k^{-5/3}$ law for the kinetic energy per unit wave number k in the low wave number part and a k^{-3} law in the high wave number part.

The vorticity conservation thus places a very important constraint upon the nature of the scale interactions. The significance of this for the kinetic energy distribution per unit wave number was first considered for the atmosphere, where observations suggested a k^{-3} dependence for a certain range of wave numbers,[161] namely for wavelengths smaller than the baroclinic instability excitation wavelength. This has been ascribed to the approximate two-dimensional nature of the motion. It should be noted, however, that the vertical velocity is not zero and that it is dynamically significant through the stretching of planetary vorticity.[162]

The properties and action of geostrophic turbulence can be extremely important for the development of the energy-containing motion in the ocean as well as for the mean circulation. Kinetic energy is favored at the expense of potential, energy may be converted from the baroclinic to the barotropic mode of motion, and transferred from the fluctuating component to the large-scale mean circulation. Energy may be carried from the baroclinic instability wavelength to barotropic modes of motion with larger wavelengths. The baroclinic energy concentrates towards eddies with a typical scale of the internal Rossby radius of deformation[163]

$$R_d = \left(\frac{\frac{g \, \Delta\rho}{\rho} h}{f^2} \right)^{1/2} = \left(\frac{g'h}{f^2} \right)^{1/2}$$

At this scale eddies above and below the thermocline may combine to produce a barotropic state of motion, permitting energy radiation towards larger scales of motion.[163]

The energy dissipation process in the ocean has not yet been clarified. It is related to three-dimensional motion and processes, but the relative role of the boundaries and the interior has not been ascertained.

For truly three-dimensional flow, vorticity is not conserved and the stretching of the vortex tubes produces a kind of energy cascade towards high wave numbers. Kolmogorov,[164] Obukhov,[165] and Onsager[166] proposed that at very large Reynolds numbers the energy fed into the motion at large scales is transferred successively to smaller scales by a cascade process. The anisotropy of the large scales of the motion is gradually wiped out and from a critical scale on the motion can be regarded as isotropic. Phillips[18] discussed the influence of stable stratification on this process. In the presence of a stable stratification there are indications that a buoyancy subrange occurs in the cascade process. The influence of the buoyancy on the turbulent motion may then cease to be important at a certain critical scale (the Ozmidov scale) determined by Ozmidov[167] and Lumley[168] and cited by Phillips[18]. Ozmidov argued that the motion in stratified conditions tends to become axisymmetric around the vertical direction, and that only components of the motion with a certain velocity shear can overcome the action of the buoyancy forces.

It is known that in locally isotropic turbulence the rate of energy dissipation per unit mass ϵ can be expressed as

$$\epsilon \propto \frac{(\Delta u)^3}{\ell} \tag{23}$$

where Δu is the velocity variation over the distance ℓ. The quantity ϵ can also be interpreted as the energy flux per unit mass through the energy cascade. Ozmidov used the relation to estimate the average velocity gradient as

$$\frac{du}{d\ell} = c \, \epsilon^{1/3} \, \ell^{-2/3} \tag{24}$$

where c is a numerical constant. On the basis of this shear a critical Richardson number can be determined

$$Ri_c = \frac{-\frac{g}{\rho_0} \frac{d\bar{\rho}}{dz}}{c^2 \, \epsilon^{2/3} \, \ell^{-4/3}} \tag{25}$$

which in turn yields, assuming $Ri_c \approx 1$,

$$\ell_c \approx \left(\frac{\epsilon}{N^3} \right)^{1/2} = \lambda_N \tag{26}$$

This is the Ozmidov scale, λ_N, where the buoyancy effects cease to be important. Subsequent studies have shown this scale to be of great significance. It may be reasonable to treat the components of motion with scales from λ_N to the dissipation scale λ_r as isotropic. In this range where inertial forces are important the Kolmogorov[164] cascade process for high Reynolds numbers should be applicable. Then the energy transfer ε through the cascade and the viscosity ν are the only significant parameters. From these the characteristic length scale for the subrange, the Kolmogorov length scale η, is obtained

$$\eta = \left(\frac{\nu^3}{\epsilon}\right)^{\frac{1}{4}} \qquad (27)$$

The typical velocity scale for the turbulence in this range is

$$v' = (\nu \, \epsilon)^{\frac{1}{4}} \qquad (28)$$

and the Reynolds number based on these scales becomes

$$\frac{v' \, \eta}{\nu} = \frac{1}{\nu} \left(\frac{\nu \, \epsilon \, \nu^3}{\epsilon}\right)^{\frac{1}{4}} = 1 \qquad (29)$$

The wave number (or dissipation wavelength) where the viscous effects become dominating is of the same order as $1/\eta$ and normally the dissipation wavelength is defined as equal to η. However, most of the dissipation occurs for smaller scales with $k \, \nu < 0.5$.[169]

In this subrange the Kolmogorov spectrum should hold, as indeed the observations by Grant et al.[51,52] showed. For very large Reynolds numbers the form of the energy spectrum is assumed to be independent of external conditions, i.e., the conditions of turbulence generation and dissipation. In this inertial subrange the spectral form is (the Kolmogorov spectrum)

$$E(k) = \text{const.} \cdot \epsilon^{2/3} \, k^{-5/3} \qquad (30)$$

where $E(k)$ is kinetic energy per unit wave number. It should be noted that the observations referred to showed this to be applicable only for scales in the range 1 to 100 cm, and that the observations were carried out in an area where the motion was characterized by a very high Reynolds number due to strong tidal flow. For a discussion of the whole turbulence spectrum reference is made to Hinze.[158]

Apart from the three scales R_d, λ_N, and η so far defined, scales can be defined which take into account the influence of the curvature and rotation of the earth. These define two spectral ranges.[170] The range influenced by curvature and rotation may be defined by the characteristic length scale

$$\lambda_\beta = \left(\frac{\epsilon}{\beta^3}\right)^{1/5} \qquad (31)$$

where $\beta = 2 \, \omega_o \cos\phi / R$ is the variation of the Coriolis parameter with latitude, with R the radius of the earth. The range being influenced by rotation and buoyancy effects may be characterized by the length scale

$$\lambda_f = \left(\frac{\epsilon}{f^3}\right)^{1/2} \qquad (32)$$

where $f = 2\omega_o \sin\phi$.

Thus a picture emerges of the turbulent motion in the sea as being characterized by a series of spectral bands. Woods[170] argued, on the basis of his flow visualization experiments,[56] that the transition between subranges appears to be abrupt and involving a distinct spectral gap, defined by the length and time scales given above. On the basis of this, Woods[170] could construct a space-time spectrum giving the location of and interactions between these distinct spectral gaps. Woods especially pointed out the possibility of direct interactions, for instance, from the baroclinic instability to the buoyancy subrange and the inertial subrange, without the energy passing through the whole range of wave numbers in the type of cascade process introduced by Kolmogorov. Implications of this are that the small-scale mixing conditions can be influenced by large-scale processes directly and that local conditions in the seasonal thermocline can be influenced by seasonal variability.

D. Boundary Layers and Stability Considerations

So far, scales characteristic for the motion and for the distribution of energy have been considered. However, in considering the forcing of the ocean it is necessary to consider the scaling of various boundary layers which can occur. The influence of the buoyancy forces on the penetration of the mechanical energy transferred from the wind may be appreciated by means of the Monin and Obukhov[171] theory. The theory was considered with special reference to the oceanic surface layer by Phillips.[18] The turbulence is supposed to be well developed and the large-scale turbulence to be characterized by a set of external parameters. These are combined to yield scales of length, velocity, and buoyancy. For the case of a steady wind with duration of one or more days the velocity scale is given by the friction velocity u_*.

$$u_* = \left(\frac{\tau_0}{\rho_0}\right)^{1/2} \tag{33}$$

where τ_0 is the wind stress on the surface.

The buoyancy flux at a level z is given by

$$B(z) = -\frac{g \cdot \overline{w'\rho'}}{\rho_0} \tag{34}$$

In the constant stress layer the conditions vary very little with depth and the quantities u* and B(O) can be combined to express the Monin-Obukhov length

$$L = \frac{u_*^3}{\kappa_0 \cdot B(O)} \tag{35}$$

where κ_0 is the von Karman constant.

The meaning of this length can be understood by considering the ratio of production of turbulent energy in a shear flow and the buoyancy term, thus

$$\frac{-\overline{u'w'}\frac{dU}{dz}}{-B(O)} = \frac{u_*^2\frac{dU}{dz}}{-B(O)} = \frac{u_*^3}{\kappa_0 z\,B(O)} = \frac{L}{z} \tag{36}$$

assuming that

$$\frac{dU}{dz} = -\frac{u_*}{\kappa_0 z} \tag{37}$$

Thus the Monin-Obukhov length represents the depth where the production term equals the buoyancy flux term. For depths much less than |L| the buoyancy flux plays a minor role but when $|z| \gg |L|$ it becomes dominant. This is the limit of strong stability which is often encountered in the sea.

For near-neutral conditions in the surface layer $L \to \infty$ and characteristic eddy sizes are proportional to z so that the length scale varies linearly with depth. The length scale function $\delta = \ell / \kappa_o z$ is then unity, which is used in the mixing length theories. In cases of strong stable stratification the eddy size is only influenced by the stratification and not by the distance from the surface. For such conditions the function $\delta \to 0$. The scale of the energy containing eddies may be defined as

$$\ell = \frac{(-\overline{w'u'})^{3/2}}{\epsilon} = \frac{u_*^3}{\epsilon} \tag{38}$$

The condition of local equilibrium then leads to the expression for the function δ as

$$\delta = \frac{1}{\kappa_o z} \frac{1}{\Phi_m (1 - Rf)} \tag{39a}$$

$$\Phi_m \equiv \frac{\kappa_o z}{u_*} \frac{\partial U}{\partial z} \tag{39b}$$

where Φ_m is a dimensionless vertical shear.

Many investigations have attempted to determine Φ_m and for stable conditions the data in general conform with the form

$$\Phi_m = (1 - c_1 \, Rf)^{-1} \tag{40a}$$

where c_1 falls in the range 4.7 to 7.0 for atmospheric turbulence data. For unstable conditions the KEYPS formula[151] is often sited

$$\Phi_m = (1 - c_2 \, Rf)^{-1/4} \tag{40b}$$

where $c_2 = 14$. The data material available for the sea is very limited, but it may be permissible to use the relations found by means of atmospheric data also for the sea.

The direct influence of wind friction is limited to the planetary boundary layer. The scale of this layer is found from an approximate balance between the Coriolis force and the frictional force due to the wind, assumed to be distributed over the whole layer. The thickness is then estimated as

$$h' \approx \frac{u_*}{f} \tag{41}$$

Near the free surface and the solid boundaries so-called wall layers develop. These are relatively thin, of the order of meters at most, and may often be disregarded.

Through the action of the wind the so-called wind-mixed layer is formed, the thickness of which is related to the balance between the mechanical energy input from the wind and the heat input. The transition to the deep water is often marked by a change in density across a pycnocline layer. This is accompanied by a change in turbulence level which is much lower in the deep water than in the wind-mixed layer. During continuous wind action the turbulence level in the surface layer is maintained and fluid is entrained across the pycnocline layer from the deep water into the surface layer. The pycnocline layer is gradually deepened at a rate called the entrainment velocity, and the density change is gradually decreased.

The entrainment process depends on the stability across the pycnocline layer and the structure of the turbulent motion. Winant and Browand[172] described the process of vortex formation when the stability is low, i.e., for small Richardson numbers. An often quoted critical number is 0.3. When the Richardson number is larger the vortex formation ceases and instead cuspy waves are formed, called Holmboe waves by Browand and Wang.[173] The amplitudes grow until the cusps are torn off and entrained into the ambient fluid. An alternative process is the Kelvin-Helmholtz instability which is coupled to the shear of the internal wave field and the mean flow. This type of instability has been observed in the seasonal thermocline.[56] Both these processes are less efficient in generating entrainment than the vortex formation process.

In boundary layer conditions the vertical transport coefficients for mass and momentum are of particular interest. For boundary layer turbulence in conditions of neutral or near-neutral density stratification the coefficients may be treated as equal in magnitude, although the ratio K_z/A_z then falls in the range 0.9 to 1.15 (e.g., Monin and Yaglom[151]). In this case the coefficients are both scaled or estimated through the expression

$$K_z, A_z = -\kappa_o u_* z \tag{42}$$

During conditions of stable stratification, however, the momentum transfer is larger than the mass transfer. According to the Monin-Obukhov theory the coefficient may then be scaled, or estimated, by the expression

$$A_z = -\kappa_o u_* L \, Rf \tag{43}$$

In these conditions the vertical shear in the layer is given by (assuming now that $U \equiv U(z)$),

$$\frac{dU}{dz} = -\frac{u_*}{\kappa_o L \, Rf} \tag{44}$$

We have here introduced the important parameter Rf, called the local flux Richardson number. This is a local parameter defined as the ratio of the local loss of turbulent energy to buoyancy forces and the turbulent energy production due to the shear, or

$$Rf = \frac{g \dfrac{\overline{w'\rho'}}{\rho_o}}{-\overline{u'w'} \dfrac{dU}{dz}} \tag{45a}$$

Strictly the flux Richardson number is defined as the ratio of the total loss to buoyancy forces and the total gain of turbulent energy from the mean flow over the whole fluid section[175], that is

$$Rf^* = \frac{\int g \dfrac{\overline{w'\rho'}}{\rho_o} dA}{\int -\overline{u'w'} \dfrac{dU}{dz} dA} \tag{45b}$$

The local flux Richardson number is the relevant parameter only when the convective transport, or diffusive flux, of turbulent energy can be neglected. This is normally done in applications.

Introducing the eddy transport coefficients

$$\overline{-u'w'} = A_z \frac{dU}{dz} \; ; \quad \overline{-w'\rho'} = K_z \frac{d\overline{\rho}}{dz} \; ;$$

$$Rf = \frac{-\frac{g}{\rho_0} \frac{d\overline{\rho}}{dz} K_z}{\left(\frac{dU}{dz}\right)^2 A_z} = \frac{K_z}{A_z} Ri \qquad\qquad (46)$$

where Ri is the Richardson number. In the limit of strong stratification and stability the flux Richardson number Rf approaches a constant value, the so-called critical flux Richardson number Rf_c. This is always less than one,[48] since loss to buoyancy is suffered through the kinetic energy of the fluctuating vertical component while viscosity effects all components. The transfer of energy between the different components of the motion is accomplished through the pressure fluctuations. Experimental evidence from laboratory investigations implies that the transfer of energy from one component to another is relatively inefficient. This led Stewart[48] to the conclusion that the maximum value of Rf (i.e., Rf_c) should be considerably less than one. Observational evidence from the sea suggests that the value of Rf_c is in the range 0.05 to 0.15.[45,106,174] This is in agreement with laboratory results.[47,134,175] The result is very important. It means that the conversion of turbulent energy into potential energy is always relatively small. Despite this the effect of buoyancy on turbulence is very significant since the production of turbulent energy is ultimately linked to the vertical velocity fluctuations.

In the limit of high stability the motion is likely to transform into a field of internal waves. We have seen that in large regions of the ocean the stability may be high, and it is of considerable interest to establish a criterium for the occurrence of turbulent motion. The stability of shear flow has been much studied[134] and we will only recall a few relevant results here. Taylor[176] showed that in the case of a continuous stratification the criterium for the existence of any system of waves was the magnitude of the Richardson number Ri. When Ri < ¼ no waves, either stable or unstable, can exist in a fluid bounded by one horizontal plane. When Ri > ¼ a series of wavelengths can exist. Taylor also showed that in the presence of viscosity the ratio $K_z/A_z < Ri^{-1}$.

Miles[177] investigated the general problem of stability of heterogeneous shear flows. He proved that sufficient (but not required) conditions for stability are that $dU/dz \neq 0$ and $Ri(z) > ¼$ in the whole fluid layer (Miles' theorem).

There is considerable observational evidence from atmospheric studies that the critical magnitude of Ri is about ¼.[151,178] Above this value the turbulence seems to become more or less suppressed. It is to be noted that this concerns fully developed flows and that Ri > ¼ is required in the whole water column. Also from the ocean there are studies indicating a critical value of Ri around ¼. It is reasonable to use Ri = ¼ as a critical value for a continuous existence of turbulence. In many areas of the ocean the mean value of Ri ≫ ¼ while Ri-values based on the layered structure, fine structure, and microstructure discussed above, may well be of the order of ¼. This seems to be in agreement with the intermittent occurrence of turbulence in relation to such layers.[56,179]

Equation 21 gives the balance between production by work of the Reynolds stress against the mean shear, the loss to buoyancy forces and the rate of viscous dissipation per unit mass. Within this simple framework the turbulence can be considered as generated by the Reynolds stress, or by gravitational collapse of Kelvin-Helmholtz instabilities or billows. Dissipation and buoyancy flux act as sinks. The Reynolds stress model can apply to zones of high shear as in the equatorial currents and the western boundary currents, whereas the Kelvin-Helmholtz model can apply to active microstructure regions.

In both cases the concept of a critical flux Richardson number Rf_c is useful when considering fairly to highly stable conditions. This implies that we consider the ratio of buoyancy flux to production as known. The energy removed from the mean flow goes in the billow case partly into mixing, partly dissipation, and is partly radiated away via internal waves.

In conditions when the gain of turbulent kinetic energy in part of the fluid cannot be neglected, the flux Richardson number should not be used. Such a condition can occur when nonturbulent fluid is entrained into turbulent fluid, whereby also the gain of turbulent kinetic energy may have to be considered in the energy balance. This led Pedersen[180] to define a bulk flux Richardson number Rf^T as the ratio of total gain of turbulent kinetic energy plus potential energy to the total turbulent energy produced corrected for the convective (diffusive) transport. The parameter Rf^T is especially relevant in conditions of weak or near-neutral stability, i.e., for low Ri numbers.

Often the densimetric Froude number is used as an alternative to the Richardson number. The definition is

$$Fr = \left[\frac{U^2}{h \frac{g \Delta \rho}{\rho_0}} \right]^{\frac{1}{2}} = \frac{U/h}{\left(\frac{g}{\rho_0} \frac{\Delta \rho}{h} \right)^{\frac{1}{2}}}$$

where h is an appropriate vertical length. We see that Fr can be regarded as equivalent to $Ri^{-\frac{1}{2}}$. The bulk flux Richardson number should be used when Fr is large, say larger than 1 (i.e., for supercritical flows).

Unfortunately it is difficult to determine Rf^T as well as Rf directly due to difficulties in determining terms in the energy equation. Studies of the terms in the equation have been carried out in the atmosphere.[181] However, so far no such relatively complete studies are available for the ocean. Nevertheless there are a few estimates of both Rf and Rf^T for conditions in the sea. Pedersen[180] found that using a constant value of $Rf^T \approx 0.1$ he could describe the variation of the vertical entrainment rate over a very large range of Ri numbers. The value of Rf often falls in the range 0.05 to 0.15.

There are also indications that for near-neutral conditions Rf decreases considerably which is in agreement with Pedersen's argument. It can be concluded that for high stabilities i.e., subcritical flows, Rf can be used with a value in the range 0.1 to 0.15, whereas for low stabilities $Rf^T \approx 0.1$ should be used. Otherwise the rate of mixing will be overestimated.

VIII. PROCESSES INFLUENCING THE MOTION

The motion in the sea is influenced by such processes at the boundaries as evaporation, precipitation, heat exchange and wind at the surface and friction at the solid boundaries. The tidal forces generate motion in the sea, and the motion will be subject to modification by the rotation of the earth. Boundary layers are generated at the surface, the bottom and the coasts, where the dynamics may be more complicated than in the interior of the sea.

A. Shelf Seas
The motion over the shelf and in the coastal boundary layer is of great interest in the present context, but it should be noted that the exchange with the open sea and the motion there will have a very marked influence on conditions in the coastal zone.

In recent years considerable attention has been given to the dynamical conditions in the coastal boundary layer and in shallow (shelf) seas in general. Some results will be discussed here in order to elucidate our present understanding.

Csanady,[182] summarizing several recent results, stated that the mean circulation in shallow sea areas arises as a residue of chaotic first-order flow episodes generated by winds, tides, and river inflow. Clearly the presence of the coastline and the bottom in combination with the influence of the Coriolis effect is also of major importance. The principal cause of the variability is the variation of the wind as one of the main driving forces. The flow recorded by local current meters is therefore highly variable whereas the flow found by means of drifters left in the water for long periods often shows a more consistent pattern. This implies that a long-term average flow pattern is present on a seasonal or longer term basis which is considerably less random than the short-term flow. This long-term flow pattern will to a large extent determine the movement of various substances, including pollutants. However, the short-term events of exchange between the coastal boundary layer and the open sea areas, generated by special wind conditions, are also of great importance for the conditions both in the coastal boundary layer and in the shallow (shelf) sea area in general. The transport capacity of one single event may be larger than the transport by the mean motion during a whole season.

The mean circulation is often interpreted as the steady state solution to the averaged Navier-Stokes equations, parameterizing the Reynolds stresses as gradient transport terms. Stommel and Leetmaa[183] studied the winter circulation over the eastern North American continental shelf by this technique. Quite apart from the requirement of empirical inputs on friction and transport parameters, Csanady[182] points out, it is not obvious that the net result of varying wind stress, tidal motion, and freshwater runoff can be expressed as a steady-state solution of the Navier-Stokes equations.

In shelf seas the main driving forces are winds and tides. Removing fluctuations which are short term in comparison with the tidal period from current meter records yields what is referred to as the first-order flow (following Csanady[182]). According to experience this flow may be modeled quite well by the equations of motion simplified by means of the hydrostatic and Boussinesq approximations. This implies that only the horizontal momentum equations need to be considered and that the internal density differences only occur in the gravitational acceleration term. In many cases the linearized version of the equations can be used neglecting the advective momentum fluxes. The momentum fluxes due to motion with shorter time scale than the tidal period appear as Reynolds stresses in the equations. Of these, however, only the components representing shear stress in horizontal planes are normally of importance. The usually large vertical gradients of these stresses are due to the wind stress and the bottom stress.

The mean or residual circulation may be defined as the flow obtained by further averaging of the current measurements over a period T long in comparison with the tidal and typical weather cycles. Often the residual current is defined as the current obtained when all tidal harmonics in the motion have been subtracted. If the first-order flow is adequately described by the linearized equations, then an average of these equations over the time T should also describe the mean circulation. The equations of the problem may then be written in the form:[182]

$$\frac{dU}{dt} - fV = -g\,\frac{d(\zeta + D)}{dx} + \frac{dF_x}{dz} \tag{47a}$$

$$\frac{dV}{dt} + fU = -g\,\frac{d(\zeta + D)}{dy} + \frac{dF_y}{dz} \tag{47b}$$

$$\frac{dU}{dx} + \frac{dV}{dy} + \frac{dW}{dz} = 0 \tag{47c}$$

Here ζ is the surface elevation and (F_x, F_y) kinematic stress in horizontal planes. The dynamic height D is defined as

$$D = \int_z^o \frac{\rho - \rho_o}{\rho_o} \, dz \tag{48}$$

where ϱ_o is reference density and $z = 0$ is the undisturbed free sea surface.

The justification for neglecting the averaged nonlinear advective terms may be investigated by means of a scaling analysis. For the first order flow u_1 with the appropriate length scale ℓ_h the magnitude of the nonlinear advection term is $u^2{}_1/\ell_h$. With an essentially one-dimensional momentum flow over the period T, the averaged term has the same order of magnitude, and it is then negligible compared to the Coriolis term if

$$\frac{u_1^2/\ell_h}{fU} = \frac{u_1}{U} \cdot \frac{u_1}{f\ell_h} \ll 1 \tag{49}$$

The velocities u_1 and U should be understood as typical amplitudes of the flows. The Rossby number of the first order flow, $u_1/f\ell_h$, is usually less than one ($\approx 10^{-1} - 10^{-2}$), implying that the linear equations may be used for the first order flow. However, the ratio u_1/U is often of the order of 10 implying that the inequality is barely fulfilled for length scales about 100 km. It should be noted, however, that the average value of $u_1{}^2/\ell_h$ in most cases will be smaller than the first order value. It may therefore be reasonable to use the linearized equations outside boundary layers; i.e., for studying the mean circulation on shelf seas without considering the coastal boundary layer.

In order to procede further the Reynolds stress terms must be parameterized. The magnitude of the kinematic Reynolds stress for the first order flow, (i.e., $F_{1x} = - \overline{u'w'}$) may during events reach a value of about 1 cm²s⁻², which is a factor of 10^2 to 10^3 less than $u^2{}_1$. This suggests that the vertical fluctuating velocity is small compared to u_1. The time scale of the motion responsible for the vertical exchange is also short compared to the inertial period; i.e., it is not significantly affected by the Coriolis effect. In such cases the turbulent kinematic stresses may be reasonably parameterized through the gradient transport hypotheses, i.e.,

$$F_{1x} = A_{1z} \frac{du_1}{dz}, \quad F_{1y} = A_{1z} \frac{dv_1}{dz} \tag{50}$$

where A_{1z} is the turbulent vertical exchange coefficient for momentum (for the first order flow). This depends upon the characteristics of the turbulence, a length scale, and a velocity scale, which both normally vary with position.

Various formulas expressing the dependence have been proposed. In shallow waters and in channel flow the length scale is proportional to the depth h and the velocity scale to the friction velocity u_*, defined by

$$u_* = \sqrt{\frac{\tau}{\rho_o}} \tag{51}$$

where τ is the appropriate boundary stress (wind stress τ_o or bottom stress τ_b). A crude approximation for both cases is[182,184,185]

$$A_{1z} = \frac{u_* h}{20} \tag{52}$$

where h is the boundary layer thickness.

In deeper waters the Coriolis effect will have an influence on the depth scale. The appropriate depth scale is then given by the thickness of the Ekman boundary layer h_E which can be given as

$$h_E = \frac{\kappa_o u_*}{f} \tag{53}$$

where H_o is the von Karman constant. This is usually given the value 0.4, but a smaller value (≈ 0.1) may be more appropriate in the present case.[182] Then the eddy coefficient becomes

$$A_{1z} = \frac{u_* h_E}{20} = \frac{\kappa_o u_*^2}{20\,f} \tag{54}$$

The formulas are matched for $h = \kappa_o u_*/f$.

These expressions do not apply close to the free or solid boundaries where other so-called wall layers occur. In the logarithmic layer the velocity profile is given by

$$\frac{du_1}{dz} = \frac{u_*}{\kappa_o z} \;,\; u_1 = \frac{u_*}{\kappa_o} \ln \frac{z}{z_o} \tag{55}$$

where z_o is a characteristic height given by

$$z_o \sim \frac{0.1\,ny}{u_*} \text{ for } d < \frac{3\,ny}{u_*} \text{ (smooth)}$$

$$z_o \sim \frac{d}{30} \text{ for } d > \frac{3\,ny}{u_*} \text{ (rough)}$$

with d the height of the roughness element. The thickness of the logarithmic layer has been found by Wimbush and Munk[49]

$$h_{1n} = \frac{2\,u_*^2}{f \cdot U_g} \tag{56}$$

where U_g is the velocity well away from the boundary. Tennekes[186] gave the alternative value

$$h_{1n} = 0.1\,h_E \tag{57}$$

Normally the logarithmic layer is order of meters thick in the surface layer and order of one meter at the bottom. A sublayer in the logarithmic layer is the constant stress layer where the stress is given by

$$\tau = \rho_o u_*^2 = \rho_o A_{1z} \frac{du_1}{dz} \tag{58}$$

In the overlapping parts of these layers the momentum transport coefficient may also be given in the form

$$A_{1z} = -\mathcal{K}_0 u_* z \tag{59}$$

For many applications in the sea the wall layers need not be considered. Then the boundary conditions are applied just outside the layers, implying that the no-slip conditions at the solid boundary do not apply.

The bottom stress for the first order flow can be expressed through the quadratic friction law

$$F_{b1x} = \rho c_b u_1 q_1 , \quad F_{b1y} = \rho c_b v_1 q_1 \tag{60}$$

where $q_1 = (u^2_1 + v^2_1)^{1/2}$ and u_1 and v_1 are velocity components just outside the bottom wall layer. The drag coefficient c_b depends upon the roughness and normally varies in the range $10^{-3} - 10^{-2}$.

The wind stress can for all practical purposes be expressed as

$$\tau_0 = c_d \rho_a W^2_{10} \tag{61}$$

where W_{10} is the wind speed at 10 m above the surface. The drag coefficient c_d can be related to the wind speed[187] as

$$c_d = (0.75 + 0.067 \, W_{10}) \cdot 10^{-3} \tag{62}$$

but often a constant value of $c_d = 1.3 \cdot 10^{-3}$ can be used. The surface friction velocity is then

$$u_* = \left(\frac{\tau_0}{\rho_0}\right)^{1/2} = \left(\frac{c_d \rho_a}{\rho_0}\right)^{1/2} \cdot W_{10} \tag{63}$$

and the momentum exchange coefficient can be directly related to the wind

$$A_{1z} = \frac{\kappa_0 c_d \rho_a}{20 \, f \rho_0} W^2_{10} \tag{64}$$

Using the numerical values $\kappa_0 = 0.1$, $c_d = 1.1 \cdot 10^{-3}$, $\varrho_a/\varrho_o = 1.23 \cdot 10^{-3}$, $f = 1.4 \cdot 10^{-4}$, one finds

$$A_{1z} = 4.8 \cdot 10^{-5} \, W^2_{10} , \quad cm^2 \quad s^{-1} , \quad W_{10} \text{ in cm} \quad s^{-1} \tag{65}$$

It may be noted that Sverdrup et al.[66] gave, for $W_{10} > 6 \, ms^{-1}$ and using $c_d = 2.6 \cdot 10^{-3}$,

$$A_{1z} = 4.3 \cdot 10^{-4} \, W^2_{10} , \quad cm^2 \quad s^{-1} , \quad W_{10} \text{ in cm} \quad s^{-1} \tag{66}$$

By means of energy considerations Kullenberg[188] found the expression

$$A_{1z} = \frac{1}{f} \left(\frac{\rho_a c_d}{\rho_0 k_W}\right)^2 \cdot W^2_{10} , \quad W_{10} \geqslant 5 \, m \quad s^{-1} \tag{67}$$

where k_w is the wind factor. With $k_w \simeq 1.8 \cdot 10^{-2}$ and the same values as above we find

$$A_{1z} = 4.1 \cdot 10^{-5} \ W_{10}^2 \ , \ cm^2 \quad s^{-1} \ , \ W_{10} \ in \ cm \quad s^{-1} \qquad (68a)$$

whereas with a wind-dependent c_d and $W_{10} = 25 \ m \cdot s^{-1}$

$$A_{1z} = 3.2 \cdot 10^{-4} \ W_{10}^2 \ , \ cm^2 \quad s^{-1} \ , \ W_{10} \ in \ cm \quad s^{-1} \qquad (68b)$$

Next comes the problem of expressing the average kinematic stress (F_x, F_y) applicable for the mean circulation. Averaging (50) we find

$$F_x = A_{1z} \frac{dU}{dz} + \overline{A'_z \frac{du'}{dz}} \qquad (69)$$

Normally $u_1 \gg U$ and $u' \approx u_1$ and the average value of the boundary stress is practically independent of the mean velocity. The average value of A_z is determined by the average stress and hence $A_{1z} \ dU/dz$ may be regarded as linear in the mean velocity. The term $\overline{A'_z \ du'/dz}$ can be considered as independent of the mean velocity. Introducing F_x, F_y in the equations (47a, b) an extra term enters in the form of

$$\frac{dc_x}{dz} = \frac{d}{dz} \left(\overline{A'_z \frac{du'}{dz}} \right), \ \frac{dc_y}{dz} = \frac{d}{dz} \left(\overline{A'_z \frac{dv'}{dz}} \right) \qquad (70)$$

These terms depend upon external variations. For instance, during storms the wind stress effect penetrates deeper than during an average wind, and the terms c_x and c_y are expressing this effect.

Csanady[182] continued to investigate the role of c_x, c_y by dividing the wind stress into components. By means of the data given by Saunders[189] on wind stress statistics Csanady found c_x and c_y to be 30 and 15%, respectively, of F_x and F_y. This illustrates that the averaging procedure may seriously mask the effect of storms. It should also be noted that our knowledge of the conditions during storms is very limited.

The bottom stress can be assumed to be almost entirely due to first order motion; i.e., the mean bottom stress is

$$F_{bx} = \rho \ c_b \ u_1 \ q_1 \ , \ F_{by} = \rho \ c_b \ v_1 \ q_1 \qquad (71)$$

The mean pressure field depends upon the surface elevation which can be determined from the continuity equation.

In many shelf seas the freshwater influx from the coastal zone due to runoff will have an influence on the circulation. We have already been considering fronts, which may also be coupled to the freshwater supply. Here we will consider, again following Csanady,[182] the influence of the first-order flow and mean circulation on the salinity gradients caused by the freshwater supply. Assuming that only offshore salinity gradients are significant the averaged salt balance equation is

$$\frac{\partial S}{\partial t} + u \frac{\partial S}{\partial x} = \frac{\partial}{\partial x} \left(K_x \frac{\partial S}{\partial x} + \overline{K'_x \frac{\partial s'}{\partial x} - \overline{u's'}} \right) \qquad (72)$$

where S is the salinity, K_x is the mean horizontal eddy exchange coefficient for salt and K'_x, u', and s' are the fluctuations determined essentially by first-order events. Taking the average over a long enough period the time-dependent term may be neglected and the depth integrated offshore salt flux becomes constant at all offshore distances x,

$$Q_s = \rho_o \int_{-H}^{o} \left(u\,S + \overline{u's'} - K_x\,\frac{\partial S}{\partial x} - \overline{K'_x\,\frac{\partial s'}{\partial x}} \right) dz \qquad (73)$$

Normally there will be a salt transport towards the coast to balance the salt deficiency of the entering freshwater. The magnitude of the terms in the equation should be considered from case to case. The advection terms are significant only when vertical salinity variations occur, and it is likely that the integrated term $u's'$ is much larger than the integrated uS. This means that salt advection through first order events dominate advection by the mean flow.

The important result is that the horizontal salt transport across the coastal boundary layer may well be caused by first order flow events, such as those generated by storms. The horizontal density gradients effecting the mean flow are then essentially independent of the mean flow.

B. The Coastal Boundary Layer

The hydrographic response to transient meteorological forcing of a semienclosed sea like the Baltic has been studied by Walin[61] who found that the baroclinic part of the motion was confined to a narrow coastal zone, which may be termed the coastal boundary layer. Walin considered time scales long compared with the inertial period $\pi/\omega_o \cdot \sin\phi$, but short compared with the time required for diffusion to change the density field. He found that the width L_c of the coastal boundary layer was given by

$$L_c \sim H_c\,\frac{N}{f}$$

where H_c is the bottom depth. The ratio of the horizontal and vertical scales of motion is in general of the order N/f for quasi-geostrophic flow. When a disturbance with a characteristic scale L acts on a system, the vertical scale of the response is expected to be $H \sim L \cdot f/N$. In many cases this scale is larger than the depth of the system implying that the response will not vary with depth, i.e., it is barotropic. This may hold for the open semi-enclosed (the Baltic) sea away from the coast. Near the coast the vertical scale is prescribed by the depth H_c and the response forced from the coast will penetrate over a distance L_c. Walin[61] found $L_c \approx 5$ km for the Baltic Sea, in fair agreement with observations.[190,191] The baroclinic disturbance in the coastal zone has a tendency to propagate along the coast in a cyclonic direction and may for certain conditions be described as a superposition of internal Kelvin waves.[61] Often strong baroclinic and jet-like coastal currents are generated by the transient meteorological forcing. These can be essential for the renewal of the coastal water mass. Csanady[192] discussed the coastal circulation on the basis of a coastal jet conceptual model. This describes an essential feature of the wind-generated circulation close to a coast in the form of a coastal jet with a width proportional to the Rossby radius of deformation. Especially the baroclinic mode becomes well-developed and has been observed in Lake Ontario as well as on the North American shelf.[192]

The important aspect of the coastal zone dynamics in relation to the transfer of pollutants is the enhanced vertical transfer which can take place in the coastal boundary layer. This is due to the combined action of the wind and the coastline implying possibilities of up- or downwelling, depending upon the wind direction. In stratified areas the vertical transport in the coastal boundary layer can be very significant compared to the vertical transfer in the open sea area.[191]

C. The Open Ocean

Only a few pertinent features will be mentioned; for a detailed discussion of dynamics reference is made to Deacon,[193] Stern,[194] and Rhines.[162] The atmospheric forcing is of dominating importance for the oceanic motion. The vertical response of the ocean can be divided into baroclinic and barotropic modes, and the horizontal response can be separated into Rossby (or planetary) waves, inertial, and gravitational waves.[162,195] Except in the near-equatorial zone the difference in frequency between Rossby and inertial waves is very large.

The unsteady oceanic motion covers a very large range of scales. For the medium- to large-scale end of the spectrum recent studies[162] have shown the importance of Rossby waves and topographic effects for the distribution of energy between the baroclinic (motion with vertical shear) and barotropic (motion without vertical shear) modes. The baroclinic energy concentrates towards eddies with a scale of the Rossby radius of deformation at which scale eddies above and below the main thermocline may combine into the barotropic mode. The energy subsequently becomes distributed to larger scales and radiates away as barotropic Rossby waves with swifter currents than the baroclinic ones.

Poleward of mid-latitudes the primary forcing by the wind falls in the frequency range between Rossby and inertial frequencies, and therefore the primary response of the ocean does not involve these waves. Despite this, linear theory has been quite successful in predicting the response in certain frequency ranges and also appears to predict the horizontal length scale of the oceanic eddies as the internal Rossby radius.

In the tropical area the atmospheric forcing is westward and Rossby waves may play a dominating role there. In the latitude belt 20 to 40° the variance of the atmospheric forcing is relatively small which may imply a relatively weak oceanic eddy field in this region.

For frequencies slightly less than the inertial one, the oceanic response is trapped near the surface, while for substantially smaller forcing frequencies the oceanic response can reach great depths. Large eastward travelling cyclones of mid and high latitudes may force an oceanic response to the bottom.

In most attempts so far, however, to explain the generation of the oceanic eddies and elucidate their influence on the mean circulation, only spatial variations of the wind field have been included. Holland and Lin[196] formulated a two-layer ocean model with a wind field varying with latitude and small horizontal momentum transport, with an eddy coefficient of $A_{x,y} = 3.3 \cdot 10^2 \mathrm{m}^2\mathrm{s}^{-1}$. They found that eddies with time and space scales similar to those observed developed spontaneously through baroclinic instability. During the development of the eddies potential energy is transformed into kinetic energy and the motion in the deep layer is intensified. The eddies move westward and the amplitude of the velocity is equal in the upper and lower layers. There is a baroclinic component in the eddy motion through which energy can be extracted from the mean density field into the eddies maintaining them against frictional dissipation. The results of Holland and Lin suggest that the eddies play an important role for the mean circulation, in particular for the flow in the deep layer. The regions of instability were in the westward flow and, less pronounced, in the eastward jet-like return flow.

The baroclinic instability mechanism is closely related to the shear in the baroclinic motion. Through this mechanism the potential energy stored in horizontal density gradients (e.g., through temperature variations) is released to kinetic energy. The instability is accomplished by motion in the direction of the horizontal temperature gradient and hence perpendicular to the current shear.[197,198]

IX. THE MAGNITUDE OF MIXING COEFFICIENTS

Here an attempt will be made to summarize present information on the magnitude of mixing coefficients, considering separately vertical and horizontal mixing. It should be emphasized that the use of mixing coefficients is a very crude approximation. However, as yet, careful use of these coefficients appears to be the best way of parameterizing effects of fluctuations on the dispersion, which we cannot otherwise take into account.

A. Vertical Mixing

For the large-scale balance, for instance considered in numerical modeling of the large-scale general circulation, the values for both vertical and horizontal mixing coefficients are derived individually from the long time (several years) average of the distribution of oceanic properties (e.g., temperature, salinity, oxygen). An alternative way is to use estimates of the rate of formation of deep water in the active zone of the Southern Ocean. Often the vertical advection-diffusion model is used[59] with

$$K_z \frac{d^2 T}{dz^2} - w \frac{dT}{dz} = 0 \qquad (74)$$

The magnitude of the ratio K_z/w is matched to a main thermocline scale (depth) of about 1 km. Knowing the upwelling velocity w in the interior of the ocean, for instance based on the rate of formation of bottom water in the polar regions, giving $w \simeq 4$ m year^{-1}, the value of $K_z \simeq 10^{-4} m^2 s^{-1}$ is found.[59,41] These values are consistent with the "classical" main thermocline theories.[199,200] It should, however, be noted that also models without vertical diffusion can explain the characteristics of the main thermocline regime.[201,202] The formulation assumes that a uniform vertical mixing in the ocean maintains the relatively uniform average distribution of properties. However, the vertical displacement, defined by $(K_z \cdot t)^{1/2}$, for a time period t equal to the advection time across an ocean basin (which is about 5 years) is only about 150 m for $K \sim 10^{-4}$ m^{2}s^{-1}. Similarly the stirring time across an ocean basin is of the order of 100 years yielding $(K_z \cdot t)^{1/2} \approx 600$ m. Both values are small compared to the depth of the ocean and the implication may be[41] that the vertical mixing is not uniform but rather takes place in active zones such as deep water formation areas, fronts, and at the ocean boundaries. The distribution of scalar properties in the interior may be determined by motion along isopycnals.[56,202]

This interpretation is supported by various studies of the mixing on a time scale of years by means of radioactive tracers. From observations of the tritium distribution in the top 1 km of the Sargasso Sea, Rooth and Östlund[81] determined mixing rates in different areas of the gyre. The local vertical mixing coefficient below the core of the 18° water was found to be $2 \cdot 10^{-5}$ m^{2}s^{-1}, assuming that vertical diffusion was the dominant factor for the development of the tritium profile below the 18° water. Observations of relatively higher concentrations of tritium at depth, at locations peripheral to the Sargasso Sea gyre, led the authors to suggest that the evolution of the tritium profile within the Sargasso Sea is not a laterally homogeneous process of downward diffusion. This may be the dominating process in the core of the gyre. It was suggested that the injection of tritium into the 18° water was almost pulse formed, leading to a downward diffusion by local processes. Since these apparently give a low transfer rate, lateral exchange or advection processes would tend to dominate the transfer of tritium to the center at greater depths, from regions where the seasonal convection extends to the corresponding water densities. The overall tritium distribution could be accounted for by a larger vertical mixing rate, corresponding to a coefficient of $7.5 \cdot 10^{-5}$ m^{2}s^{-1}.

Rooth and Östlund[81] also found that a lateral mixing coefficient of 10^4 m²s⁻¹ would be required to negate the significance of local vertical mixing corresponding to $K_z = 2.10^{-5}$ m²s⁻¹. Tritium profiles observed in the North Pacific Ocean are also consistent with a vertical mixing coefficient of 2.10^{-5} m²s⁻¹.[203]

By means of tritium and radiocarbon observations Peterson and Rooth[204] estimated the time scale for deep convective mixing in the Greenland and Norwegian Sea to 30 years, implying a mixing rate of 100 m per year. This result suggests that the mixing can be at least an order of magnitude larger in high latitudes than in the center of the subtropical gyre. The data given by Östlund et al.[86] show tritium extending in a tongue into the Sargasso Sea between the 10° and 15° isopycnals and along the bottom south to about 37°N (see Figure 10). The diffusion regime in the 10° to 18° water in the Sargasso Sea is evident in plots of tritium concentration vs. potential temperature. The authors concluded that the primary transport mechanism from the high latitudes was large-scale eddies.

Jenkins and Clarke[205] used observations of the He-3 distribution to study the interior mixing. Along the western boundary of the North Atlantic they found a tongue of excess He-3 extending from about 51°N to the equator. The source was considered to be a fracture zone around 51°N, 30°W. In the South Atlantic a tongue of He-3 was observed extending from the Antarctic northwards at depths between 1 and 2 km. For the latter case the authors assumed steady state and a balance between vertical diffusion and horizontal meridional advection and found the ratio K_z/V to be in the range $(5.8 \text{ to } 0.6) \cdot 10^{-2}$ m, decreasing northwards (V is the advection velocity). These values conform with other results and they suggest that relatively marked vertical mixing occurs in the southern region around 45°S.

Rather than using the large-scale distribution of properties, direct observations of the microstructure distribution can be used to determine vertical mixing rates using Equations 2 and 3, as done by Osborn and Cox.[92] The Cox number Co is thereby determined from the microstructure observations.

Using this technique K_{zT} has been estimated for different regions of the ocean. For the open Pacific values of $K_{zT} \sim 10^{-6}$ m²s⁻¹ were found.[98,206] In the California current and above the main thermocline in the Gulf Stream the average values were of the order of 10^{-3} m²s⁻¹.[96,207] The open ocean values are clearly less than those obtained from the average distributions whereas the values in more active zones are larger. The balance assumed in the microstructure model neglects the horizontal production terms which may well be important, at least in coastal zones.

Schmitt and Evans[208] used salinity profiles, observed in open ocean regions where saltfingers may occur, to calculate an equivalent vertical exchange coefficient for salt under the assumption that the transporting mechanism was saltfinger convection. They argued that saltfingers can be present in large areas of the open ocean on the basis of the observation that in the central parts of the upper ocean the ratio

$$\alpha \frac{\partial T}{\partial z} \Big/ \beta_s \frac{\partial S}{\partial z} \approx 2 \tag{75}$$

which fulfills the criterium[127] for the occurrence of saltfingers

$$\alpha \frac{\partial T}{\partial z} \Big/ \beta_s \frac{\partial S}{\partial z} < \frac{\mathcal{K}}{\mathcal{K}_s} \tag{76}$$

or for the case of a sharp interface[209]

$$\alpha \frac{\partial T}{\partial z} \Big/ \beta_s \frac{\partial S}{\partial z} < \left(\frac{\mathcal{H}}{\mathcal{H}_s} \right)^{3/2} \tag{77}$$

Furthermore the fingers are not destroyed by the presence of shear[129] but seem to be aligned into rolls with no apparent change of the fluxes for Richardson numbers down to about 6, maybe even to the Kelvin-Helmholtz instability can occur.[210] The salt flux can be calculated from observations of the salinity changes across regions of increased salinity gradients and estimates of the fraction of the water column that is affected by saltfingers, which the authors assumed to be the areas of high gradients. The fraction gives an intermittency factor for the flux calculation, similar to the one used by Thorpe[179] and also related to the intermittency factor given by Woods and Wiley.[91] The buoyancy flux due to salt transport at a fingering interface is given as

$$\beta_s F_S = a \cdot (g \cdot \mathcal{H})^{1/3} \cdot (\beta_s \Delta S)^{4/3} \tag{78}$$

where a is a constant which for

$$\alpha \frac{dT}{dz} \Big/ \beta_s \frac{dS}{dz} < 2$$

has been determined to 0.1 in laboratory experiments. The laboratory observations in general shows good agreement with the 4/3 law (e.g., Schmitt[211]). The equivalent transport coefficient was by Schmitt and Evans[208] determined from the ratio

$$IF_s \beta_s \Big/ \beta_s \overline{\frac{dS}{dz}}$$

where I is the intermittency factor. Calculations were carried out for several stations in the western subtropical North Atlantic Ocean, for the depth range 200 to 1000 m. The corresponding values of K_{zs} in the range $(5 \text{ to } 80) \cdot 10^{-6} \mathrm{m^2 s^{-1}}$, correspond well with other results. Thus there are clear indications that saltfinger convection is an important transport process in some stably stratified parts of the ocean. Areas where it can occur are in the Mediterranean, in the Mediterranean outflow,[212] and also in the open sea.[208]

For the surface layer of the ocean vertical mixing rates have been directly observed by means of artificial tracers (or by using the seasonal penetration of heat). The values vary considerably with time and region under consideration. Dye experiments have given values in the range $(1 \text{ to } 100) \cdot 10^{-4} \mathrm{m^2 s^{-1}}$ for the layer above seasonal thermoclines. These experiments cover time scales from hours to weeks and space scales from several tens of meters to several tens of kilometers.[5,105,106] The mixing rates appear to vary with the wind and the stratification, and they are generally larger than below the thermocline. The average seasonal (or monthly) mixing rates determined by means of the heat wave appear to be very similar. The values found by Matthäus[214] for the thermo- and halocline layers in the central Baltic Sea generally agree very well with the results of Kullenberg.[83,106]

The vertical mixing in and below pycnocline layers has been investigated by means of dye studies. The visual observations by Woods[56] in seasonal thermoclines demonstrated the intermittency of the mixing with relatively intense mixing periods interrupted by almost laminar flow conditions. The mixing occurred in the form of billows, generated by shear instabilities on vertical scales of a few tens of centimeters. During the intense periods the mixing rates correspond to $K_z \sim 2 \cdot 10^{-5}$ m²s⁻¹. Clearly the intermittency implies that the average values of K_z are much lower than the value found from the vertical advection-diffusion balance. Woods[136] estimated that the mixing periods covered about 10% of the time and that the Reynolds number during these periods was of the order of 10^3 in the mixing billows. Woods' observations were on scales of about 100 cm and some hours. Subsequently billows have been observed in seasonal thermoclines in open ocean areas as well as in lakes.[179] The scale of the billows appears to conform with the interpretation that they are due to shear instabilities.

Clearly the efficiency of the mixing driven by intermittent occurrence of local small-scale turbulence is low. Woods[215] argues that the main transfer into and across the thermocline layer was due to mixing along isopycnals with efficient transfer in areas of high baroclinicity. The implication is that the mixing would be influenced by relatively large scale fluctuating motion, like meso-scale eddies, influencing the slope of the isopycnals over several tens of kilometers.

Woods, Wiley, and Briscoe[70] suggested that the vertical transport in the seasonal thermocline is controlled by undulation of unstable frontal zones. Such a process would lead to a vertical diffusion coefficient about $(1 \text{ to } 3) \cdot 10^{-4}$ m²s⁻¹.[215] It would also be an intermittent process giving rise to what might be called bursts of mixing. However, mixing in frontal zones has also been related to interleaving and double-diffusive phenomena[146] as well as to cabbeling[216] in zones with marked temperature and density variations.

Kullenberg[83,105,106] used observations of pulse-formed dye layers (compare Figures 14 and 15) inside the pycnocline layer to determine vertical mixing coefficients, for time scales up to about 1 day. Dye layers with sharp boundaries could be identified by their thickness, usually in the range 30 to 100 cm, and their position in the temperature structure. Vertical mixing coefficients were determined from experiments covering a large range of different conditions in the Baltic Sea, the North Sea, and the Mediterranean, as well as a series of fjords. Values in the range $2 \cdot 10^{-6}$ to $4 \cdot 10^{-5}$ m²s⁻¹ were found in the thermo- and halocline layers in open sea areas at depths from 20 to 70 m.

Direct observations of the fluctuating parts of the velocity, temperature, and salinity fields with microsensors mounted on towed bodies have also been used to study the mixing.[51,52,54,217,218] Nasmyth[54] determined vertical mixing coefficients for heat in the range $(0.06 \text{ to } 3.1) \cdot 10^{-4}$ m²s⁻¹ from observations off British Columbia at depths ranging from the thermocline to 330 m. Gibson et al.[218] used observations in the Pacific equatorial zone at depths just above the maximum velocity core in the Undercurrent and found $K_{zT} \approx 0.5 \cdot 10^{-4}$ m²s⁻¹ at 1°N and $27 \cdot 10^{-4}$ m²s⁻¹ at equator.

Osborn[219] used the simplified turbulent energy equation (21) and the assumption of a constant critical flux Richardson number $Rf_c = 0.15$ to estimate vertical, or cross-isopycnal mixing rates for density from measurements of local dissipation rates. We have the following relations

$$K_z \frac{d\bar{\rho}}{dz} = -\overline{w'\rho'} \, , \, \frac{g}{\rho_0} \frac{\overline{w'\rho'}}{-\overline{u'w'} \frac{dU}{dz}} = 0.15 \qquad (79a,b)$$

which in Equation 21 give

$$K_z = \frac{0.15\epsilon}{0.85N^2} = 0.2 \frac{\epsilon}{N^2} \qquad (80)$$

Using observed dissipation rates and stratification parameters, K_z was determined and compared with other independent calculations for temperature, using microstructure observations and the Osborn and Cox[92] model. Observations from the Atlantic Equatorial Undercurrent and an area adjacent to the Azores were used. In the former area, the agreement between the methods was good in the region of the high shear above the core of the Undercurrent with values about $4 \cdot 10^{-6}$ m^2s^{-1}. The agreement was reasonable in the core region with values about $2 \cdot 10^{-6}$ m^2s^{-1} but less satisfactory in the region below the core where the values varied over the large range $4 \cdot 10^{-6}$ to $3 \cdot 10^{-4}$ m^2s^{-1}.

In the area adjacent to the Azores the mixing rates varied not only with depth but also with distance from the islands. The range was $3 \cdot 10^{-5}$ to $5 \cdot 10^{-4}$ m^2s^{-1}. Finally, Osborn used data collected from other areas in the North Atlantic showing the variation of the mixing rates for the seasonal and main thermoclines.

The vertical mixing in the bottom boundary layer of the ocean interior has been investigated using observed profiles of various tracers. Sarmiento et al.[220] calculated K_z from profiles of Rn-222 and Ra-228 over a layer ranging from 50 to 1500 m from the bottom. Although the range of K_z was large ($(5$ to $400) \cdot 10^{-4}$ m^2s^{-1}), the majority of values were in the range $(10$ to $50) \cdot 10^{-4}$ m^2s^{-1}. The authors found that the values of K_z were proportional to N^{-2}, using the potential density, with the best fit being

$$K_z = 4 \cdot 10^{-6} \cdot N^{-2}, cm^2 \ s^{-1} \tag{81}$$

This result agrees with the model proposed by Welander[221] based on the assumption of a constant energy flux through the boundary layer.

Armi[222] presented vertical profiles of salinity, potential temperature, and light scattering which suggested that the main cross-isopycnal mixing occurred within 50- to 150-m thick layers at the ocean boundaries. From these boundary layers the properties then become advected along the appropriate isopycnal surface into the interior. The vertical boundary layer mixing was estimated as (compare Equation 14)

$$K_z \approx \frac{1}{15} u_* h, \ u_* \approx \frac{1}{30} U \tag{82}$$

where u_* is the friction velocity and h the layer thickness. With $\bar{h} \approx 50$ m and $U \approx 0.1$ ms^{-1}, Armi[222] found K_z (boundary) $\approx 10^{-2}$ m^2s^{-1}. Assuming that the flux in the boundary layer served the entire basin and equating the boundary flux with an apparent interior flux Armi arrived at K_z (interior) $\approx 10^{-4}$ m^2s^{-1}. He concluded that boundary layer mixing may well account for the vertical flux in the deep ocean. Results of a similar nature for semi-enclosed areas like fjords and the Baltic Sea have been given by Stigebrandt[174] and Shaffer.[62]

The importance of the mixing in the boundary layer and along topographic features for the distribution of oceanic properties was further discussed by Armi.[223] He advanced the hypothesis that spatial variations in the eddy diffusivity may be responsible for some features of the observed distribution of oceanic scalars as an alternative to the advective-diffusive interpretation. The spatial variation of the mixing is coupled to the variability of the oceanic eddy field. In regions where the gradient of the eddy diffusivity is at least of the same order as the mean advection the diffusivity variation could play a role. Armi considers the Mediterranean outflow as an example where this may apply.

It may be concluded that the interior vertical mixing is weak and that more pronounced vertical mixing occurs in high latitudes, at the equator, along topographic features (ridges, seamounts), in upwelling areas, and at the ocean bottom boundaries. The results are summarized in Table 2.

Table 2
SUMMARY OF RESULTS ON MAGNITUDE OF THE
VERTICAL MIXING COEFFICIENT K_z

Method of determination and region of value	$K_z \cdot 10^4 \, \text{m}^2 \cdot \text{s}^{-1}$
Circulation model, oceanic tracer distribution	1—1.5
Mediterranean outflow, Atlantic intermediate water	0.35—0.7
Tritium fallout, Sargasso Sea	0.2
Tritium fallout, Greenland and Norwegian Seas	3
Double diffusion, Cox Number, open Pacific	0.01
California current	10
Gulf Stream	10
Saltfinger flux, Atlantic, central	0.05—0.8
Dye mixing (tracing), surface layer	1—200
Dye visual (divers), thermoclines	0.2
Dye diffusion (tracing), pycnoclines	0.02—0.4
Towed micro sensors, Pacific, coastal zone	0.06—3
Pacific equatorial	0.5—27
Radon and radium profiles, Atlantic bottom boundary	10—50
Dissipation measurements, Atlantic Equatorial Undercurrent	0.02—3
Dissipation measurements, Atlantic near the Azores	0.3—5

B. Horizontal Mixing

We turn now to the magnitude of the horizontal mixing. This is known to be scale-dependent and many investigations suggest the relation (Figure 18)

$$K_h \propto \ell_h^n \quad 1.1 \leqslant n \leqslant 1.4 \tag{83}$$

Very often the value n = 4/3 is given which is equivalent to the Richardson law.[224] This result can be obtained from the similarity hypotheses.[225] For large diffusion times, implying that $\overline{(r^2)}^{1/2} \gg r_o$ where r and r_o are the separation distances at times t and t = 0, respectively, Batchelor found

$$\frac{1}{2} \frac{d}{dt} \overline{r^2} \propto \epsilon^{1/3} \cdot \overline{r^2}^{2/3} \Rightarrow K_h \propto \epsilon^{1/3} \cdot r^{4/3} \tag{84}$$

For small diffusion times Batchelor derived

$$\frac{1}{2} \frac{\overline{dr^2}}{dt} \propto \epsilon^{2/3} \cdot r_o^{2/3} \cdot t \tag{85}$$

On the basis of dye diffusion studies in the near-surface layer covering time periods up to a month and length scales in the range 30 to 100 km, Okubo[73,226] found that empirically over the whole range the scale dependence could be expressed as (compare Figures 18 and 19) (σ^2_{rC} and t in cm² and s; K_h and ℓ_h in cm²s⁻¹ and cm)

$$\sigma^2_{rC} = 0.0108 \cdot t^{2.34} \tag{86a}$$

$$K_h = 0.0103 \, \ell_h^{1.15} \tag{86b}$$

Here σ^2_{rC} is the variance of the rotationally symmetrical concentration distribution obtained by transforming the real (observed) distribution by conserving areas enclosed by isolines (for definitions see this volume Chapter 2). The similarity hypotheses imply

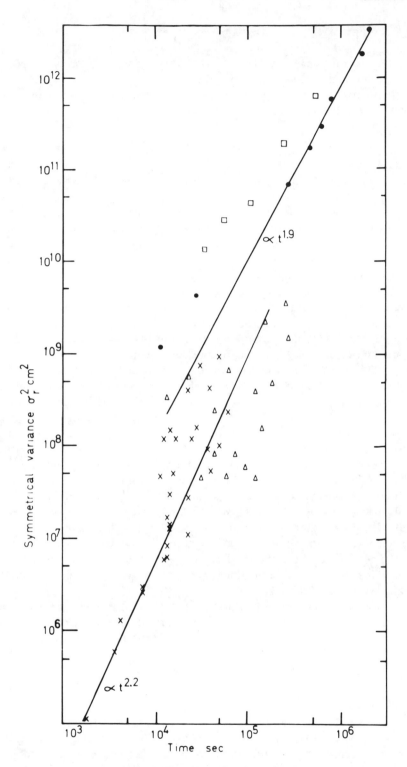

FIGURE 18.   Time dependence of horizontal and radial (symmetrical) variance of dye concentration distribution; x, observations in various coastal areas; △, observations in Lake Ontario; •, observations in central North Sea; □, observations in southern North Sea. (From Kullenberg, G., An Experimental and Theoretical Investigation of the Turbulent Diffusion in the Upper Layer of the Sea, Vol. 25, Institute of Physical Oceanography, University of Copenhagen, 1974.)

$$K_h = \text{const.} \cdot \epsilon^{1/3} \cdot \varrho^{4/3}, \sigma_{rc}^2 = \text{const.} \cdot \epsilon \cdot t^3 \qquad (87)$$

However, the assumptions underlying the similarity theory are clearly not fulfilled for the oceanic motion over these ranges of scales.

Dye experiments carried out in pycnocline layers over the limited scale range 10 m to 1 km gave the relation[105]

$$K_h = 1.3 \cdot 10^{-3} \cdot \varrho_h^{1.3} \qquad (88)$$

implying $K_h = 0.4$ m²s⁻¹ for a scale of 1 km whereas the surface data give $K_h = 0.6$ m²s⁻¹.

The results presented in Okubo's diagrams suggest that in the open ocean the diffusion can increase with scale up to about 10⁴ km. In limited or semienclosed areas it may be expected that the diffusion would approach a constant value beyond certain scales. Some evidence for this to be the case has been presented by Talbot and Talbot[227] for the southern North Sea and by Kullenberg,[228] combining experimental results from the Baltic Sea, the North Sea, and the Mediterranean.

Observed long-time average distributions of natural tracers in the ocean have been used in numerical models to determine the rates of mixing giving the best fit to the distributions, when used in combination with the Stommel and Arons[79,229] general (abyssal) circulation model. Surprisingly consistent results have been obtained with K_h in the range (5 to 10)·10² m²s⁻¹ (see the review by Veronis[230]). The corresponding deep upwelling velocity is about 5 m year⁻¹, which for the vertical models would imply a uniform vertical mixing rate of $K_z = 1.5 \cdot 10^{-4}$ m²s. The Mediterranean outflow has been used by several authors to investigate the mixing at intermediate levels in the central Atlantic. Needler and Heath[231] found vertical mixing values in the range (0.35 to 0.7)·10⁻⁴ m²s⁻¹ and horizontal in the range (1.5 to 3.5)·10³ m²s⁻¹ for typical known advection velocities. The authors concluded that diffusion was not an important process for the overall dynamics of the main thermocline. Similar results were obtained by Richardson and Mooney.[232] For the Mediterranean outflow the surface layer dye results would predict $K_h \sim 6 \cdot 10^3$ m²s⁻¹ which clearly is too large. This shows that the mixing rates are lower at intermediate depths than in the surface layer.

In the meso-scale range the mixing has been studied by means of Gulf Stream rings.[233,234] The latter authors[234] considered $K_h \sim 10$ m²s⁻¹ and $K_z \sim 0$ to be the most reasonable values for describing the time-development over 1 to 2 months, on the basis of an axially symmetrical model.

An alternative way to express the horizontal mixing is through a diffusion velocity. The mean diffusion velocity P defined by the theory of Joseph and Sendner[235] is commonly used. The concentration distribution in the rotationally symmetrical form is

$$C(r,t) = \frac{M}{2\pi (Pt)^2 h} e^{-r/Pt} \qquad (89)$$

where r is the distance from the drifting point of injection, corresponding to the equivalent radius, M is the diffusing mass, and h is the thickness. The P values reported from dye experiments in the surface layer fall in the range (0.5 to 1.5)·10⁻² ms⁻¹ while values representative for subsurface layers are considerably lower, in the range (0.05 to 0.2)·10⁻² ms⁻¹. The deep experiment by Schuert[236] at 300 m in the Pacific gave P ~ 0.1·10⁻² ms⁻¹. Ewart and Bendiner[237] found a slightly lower value of 1000 m depth. It should be noted that the diffusion velocity is one to two orders of magnitude less than the advection velocity. It should also be noted that the rotational symmetry is not

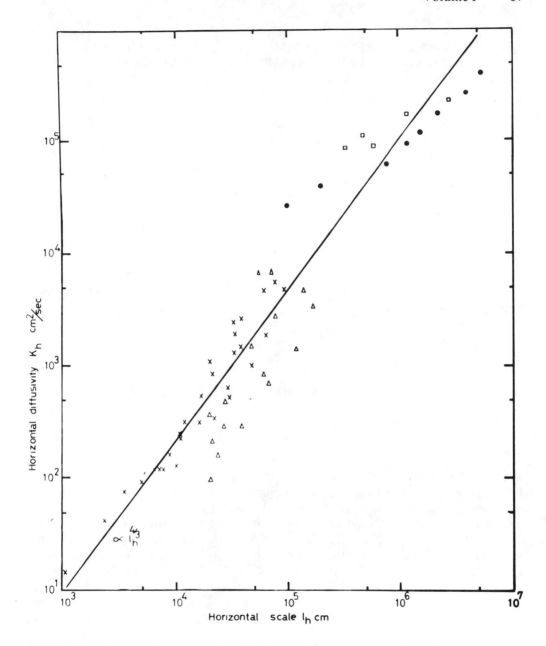

FIGURE 19. Horizontal turbulent (eddy) diffusion coefficient as determined from dye experiments vs. length scale; same notations and data set as in Figure 18. (From Kullenberg, G., An Experimental and Theoretical Investigation of the Turbulent Diffusion in the Upper Layer of the Sea, Vol. 25, Institute for Physical Oceanography, University of Copenhagen, 1974.)

normally found in observations, mainly due to shears in the motion. Nevertheless, this approach is often used. Results on the horizontal mixing are summarized in Table 3.

C. Ratio of Vertical to Horizontal Mixing

It is of interest to consider the ratio K_z/K_h. From the above values of K_z and K_h the range of the ratio is found to be $(3 \text{ to } 300) \cdot 10^{-9}$. In a region where the isopycnals have a slope i, a truly horizontal mixing rate given by K_h leads to a vertical or nearly cross-isopycnal transport rate given by

Table 3

SUMMARY OF RESULTS ON MAGNITUDE OF HORIZONTAL
MIXING COEFFICIENT K_h AND DIFFUSION VELOCITY P

Method of determination and region of value	K_h (m^2 s^{-1})	P (cm s^{-1})
Circulation model, oceanic tracer distribution	$(5—10) \cdot 10^2$	—
Mediterranean outflow, Atlantic intermediate water	$(1.5—3.5) \cdot 10^3$	—
Gulf Stream ring	$0.1—10$	—
Dye diffusion, large scale, surface layer	$10^3—10^4$	—
Dye diffusion, small scale, pycnoclines	$10^{-1}—10^0$	$0.05—0.4$
Dye diffusion, surface layer		$0.5—1.5$
Dye Diffusion, Pacific intermediate water		$0.04—0.13$

$$K_z = K_h \cdot i^2 \qquad (90)$$

This may be seen as follows. The ratio of vertical to horizontal transports is

$$\frac{Q_z}{Q_h} = \frac{K_z \dfrac{dT}{dz}}{K_h \dfrac{dT}{dx}} = \frac{K_z}{K_h} \cdot \frac{1}{i} \qquad (91)$$

The cross-isopycnal flux $Q_{cr} = Q_h \cdot i$ which together with $Q_{cr} \approx Q_z$ leads to

$$\frac{Q_{cr}}{Q_h} = i \simeq \frac{Q_z}{Q_h} = \frac{K_z}{K_h} \cdot \frac{1}{i} \quad \text{or} \quad \frac{K_z}{K_h} = i^2 \qquad (92)$$

The result would imply that the isopycnals have slopes in the range $5 \cdot 10^{-5} \leqslant i \leqslant 5 \cdot 10^{-4}$, using the above values. Observations of the mean density distribution along north-south sections in the Atlantic Ocean[74] show isopycnal slopes in the depth interval 200 to 2000 m varying considerably with latitude. In high latitudes, 50 to 60°, the slope is of the order 10^{-3}; in the latitude belt 50 to 20° it is about $2 \cdot 10^{-4}$, and in the equatorial zone 10°N to 10°S the slope is about $3 \cdot 10^{-5}$. These values fall fairly well within the range expected from the relation $K_z/K_h = i^2$. The result also supports the idea that the vertical mixing rates are not uniform throughout the ocean. It seems clear that vertical mixing is intense in certain high-latitude regions of deep water formation where the isopycnals also have a large slope. The vertical mixing is also intensified in areas of up- or downwelling. In the interior and across pycnocline layers the vertical mixing is relatively weak. The vertical mixing in the ocean is probably everywhere characterized by intermittency although the intermittency factor may vary. The horizontal mixing is probably more uniform but is also intermittent, and clearly varies with scale.

The flow as well as the mixing in the ocean to a large extent takes place along isopycnic surfaces. This may also be the case for coastal areas and semienclosed seas. It may therefore be more appropriate to discuss the mixing in relation to such surfaces rather than considering vertical and horizontal mixing. The distributions and transfers could then be described in salinity-temperature space.[238] Formally we may consider horizontal mixing to take place approximately along isopycnals. However, the implication that the corresponding vertical mixing is approximately given by $K_h \cdot i^2$ should be used with great care.

X. ENERGY SOURCES AND MIXING PROCESSES

There are several possible energy sources for mixing in the ocean: (1) tidal motion, (2) cooling, evaporation, and radiation, (3) mechanical energy transferred from the wind, (4) thermo-haline energy released through molecular processes such as double diffusion, (5) various instabilities (e.g., the baroclinic instability converting potential energy into kinetic energy) and shear instabilities, and (6) breaking waves, both at the surface and internally.

Munk[59] showed that the total dissipation of tidal energy per unit time and the energy required for the known oceanic mixing rates are very similar. It has been estimated that about 30 to 40% of the tidal energy is dissipated by bottom friction in shallow seas. The influence of tidal mixing on the conditions and generation of fronts in shelf seas has been clearly demonstrated.[239] Internal motion in the ocean can be generated by pumping from tides along the continental slopes.[240] The semidiurnal internal tide in the ocean contains energy equal to 10 to 50% of the surface tide,[240] and dissipation of tidal energy by internal tides could be the main energy source for mixing in the interior. The internal waves generated by tidal action across the sills in certain fjords have been shown to be important energy sources for the deep water mixing in such fjords.[174]

The double-diffusive process as discussed above appeared to be important for the transfer of heat and salt and for releasing potential energy. Other thermohaline effects are due to cooling and evaporation. These factors together with the wind energy are instrumental in breaking down the seasonal thermoclines and also act in the formation of deep waters. Thermohaline processes also occur as a result of ice formation since salt is released thereby. The significance of this process is not known but it is quite possible that it is important. The baroclinic instability process appears to be very important in converting energy received in the surface layer to kinetic energy in the deep water. It is, however, not established how significant the process is although there is evidence in favor of it.

The energy transferred from the wind is not only a main driving force for the circulation but also for the mixing in the surface layer. The total energy flux from the wind per unit area may be written in the form

$$E = k_W \tau_0 W_{10} \tag{93}$$

where k_W is the wind factor and W_{10} the wind speed at 10 m above the surface. The surface wind-generated current V_o has been found empirically to be proportional to the wind speed, i.e.,

$$V_O = k_W W_{10} \tag{94}$$

with k_W normally falling in the range $(2 \text{ to } 4) \cdot 10^{-2}$. Sometimes it is pertinent to consider the whole surface mixed layer to be moving as a slab with a velocity proportional to wind speed, using a wind factor which is about one tenth of the value used for the surface current V_o. Considerable observational evidence shows that only a small part of the energy transferred from the wind is consumed for vertical mixing.[241,242] Most of the energy goes into kinetic energy and wave energy and normally only about 10% is used for changing the potential energy of the water column. The total wind energy per unit time and area may be written

$$E_W = \tau_0 W_{10} \tag{95}$$

and the ratio of total energy consumed for vertical mixing to the total wind energy may be given as

$$R_1 = \frac{\int_{-h}^{0} K_z N^2 dz}{\tau_0 W_{10}} \qquad (96)$$

This ratio has been found to vary in the range $(1 \text{ to } 3) \cdot 10^{-3}$ (e.g., Kullenberg[188]). Observations obtained during the passage of strong atmospheric fronts with velocities reaching storm force, show that very efficient deepening of thermoclines can occur during such conditions.[243-245]

The wind-generated waves contain a considerable amount of energy and no doubt are a major energy source. Available evidence suggests, however, that the energy released through the breaking of waves does not penetrate deep but only influences the top meters.[246] It is evident, however, that wave energy plays a dominant role in the near-shore zone. Internal waves also transfer energy which is particularly important for mixing in the deep waters where internal waves may be the only energy source. Internal waves are undoubtly important for the mixing in deep water both by breaking in the interior and by breaking along sloping boundaries.

The Langmuir circulation, called so after its discoverer Langmuir,[247] consists of fairly narrow convergence zones with narrow bands of downward motion, aligned within a few degrees alongwind, separated by divergence zones with upward motion over a wider area (Figure 20). This organized surface layer circulation is very often observed for wind velocities above 3 ms^{-1}, and is manifested by streaks of foam or floating debris formed by the convergent motion. The spacing of the streaks vary considerably, from meters to several hundred meters. In stable (heating) conditions the vertical circulation appears to be limited to the upper 10 m or less.

Although no theory exists which can explain all the characteristics of the Langmuir circulation in a satisfactory way there seems to be general agreement that the circulation is wind generated, possibly indirectly through the action of the waves.[248] The response to a change of the wind direction is rapid. Langmuir circulation can be an important mixing mechanism in the surface layer, and may control the downward diffusion of momentum transferred from the wind to the water. It appears, however, that the mixing does not penetrate very deeply, possibly not through the whole mixed layer.[248]

The mixing process is often related to the energy source and different processes dominate in different regions of the ocean. It has been suggested that the mixing essentially occurs along isopycnals. In regions where these are approximately horizontal, mixing along them will give no contribution to vertical mixing. Breaking waves, billows, and shear instabilities will then generate vertical mixing. In areas where the isopycnals have an appreciable slope, mixing along them will also give rise to vertical mixing.

Mixing along isopycnals, or in essence horizontal mixing, seems to be mainly caused by mesoscale eddies. These stir the ocean and smaller scales of motion are required to fill the gap down to molecular diffusion; these can be internal waves, instabilities, interleafing, or double diffusion.

In many cases horizontal and vertical mixing is coupled. Mixing along sloping isopycnic surfaces will give rise to an apparent vertical and horizontal mixing, and vertical mixing combined with vertical shear will generate an apparent horizontal mixing. Generally speaking, mixing in the direction of a velocity gradient will, in combination with the gradient, generate an apparent mixing in the direction of the flow. This is the

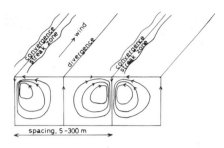

FIGURE 20. Schematic distribution of motion in Langmuir circulation cells. (From Pollard, R. T., *A Voyage of Discovery,* Angel, M., Ed., Pergamon Press, New York, 1977, 235. With permission.)

concept of shear diffusion introduced by Taylor.[249,250] It is an important mixing mechanism in many areas for scales up to several tens of kilometers. In particular, the vertical shear effect appears to be important. This effect cannot be included explicitly in a one-layer vertically integrated model. The effect may be included implicitly by enhancing the horizontal diffusion coefficient K_h by interpreting the shear effect as generating an apparent horizontal mixing. It turns out that for time periods long compared to the vertical mixing time $t_o = h^2/K_z$ (h being the thickness of the layer) it is a reasonable approximation, but for time periods shorter than t_o this is not the case; then the shear effect is much stronger than the diffusion. For long times $t > t_o$ the effects are additive linearly, both giving rise to a concentration variance $\sigma^2_{rh} \propto t$, but for short times $t < t_o$ the shear effect yields a variance $\sigma^2_{rh} \propto t^3$. This suggests that two different cases can be considered, namely the vertically bounded or long-time case, $t < h^2/K_z$, and the vertically unbounded or short-time case.

Various models for the shear effect have been formulated and to a certain extent compared with observations.[251-255] A review of these models was given by Kullenberg.[106] In general the models assume a simplified form (often constant) of the vertical shear, include a steady and a time-dependent oscillating current, and predict the apparent horizontal mixing. Pure horizontal diffusion, as expressed by K_x, K_y, may or may not be included in the model, but K_z is always present and vertical advection is neglected. Kullenberg[106] included horizontal oscillations in two dimensions:

$$\frac{dC}{dt} + u\frac{dC}{dx} + v\frac{dC}{dy} = K_z\frac{d^2C}{dz^2} \tag{97}$$

In this model the apparent horizontal mixing is interpreted as an effect of the combined action of advection and vertical diffusion. The horizontal current field is assumed to consist of mainly low-frequency, relatively large-scale motions creating primarily an advective type of transport. In the model this is specified as a mean current varying with depth only and one superimposed oscillating current component. The amplitude varies with depth but the period as well as the phase are treated as independent of depth. The model yields the concentration variance for a line source, neglecting oscillating components

$$\sigma^2_x = \frac{2}{3} K_z \left(\frac{dU_o}{dz}\right)^2 t^3 + \left(\frac{du_o}{dz}\right)^2 \frac{K_z t}{\omega^2} \tag{98a}$$

$$\sigma^2_y = \left(\frac{dv_o}{dz}\right)^2 \frac{K_z t}{\omega^2} \tag{98b}$$

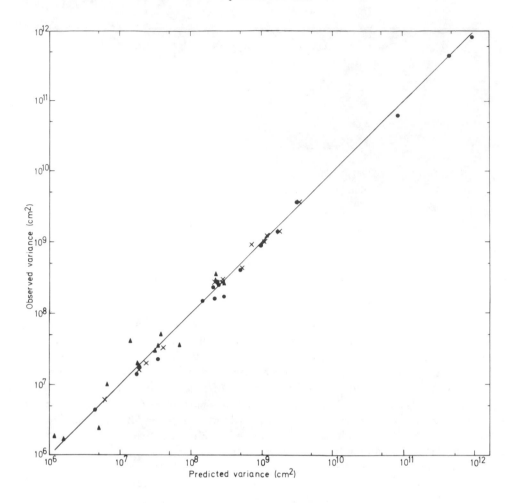

FIGURE 21.   Observed (dye tracing data) vs. predicted horizontal concentration variances using shear diffusion model; •, longitudinal variance; △, transverse variance; x, total variance. (From Kullenberg, G., *Rapp, P. V. Reun. Cons. Int. Explor. Mer,* 167, 86, 1974. With permission.)

where U_o, u_o, and v_o are the current velocities in the center of the diffusing layer, and the variances are in the x, y, and z directions, respectively. It is noted that all terms are proportional to the vertical mixing as expressed by K_z, and that the effect due to the steady shear dominates. The lateral variance σ^2_y is determined by the oscillating current component and large periods become increasingly important with increasing diffusion time.

For $K_z = 0$ the variances become zero. However, for a source with finite thickness this is not the case. Integration over the initial thickness yields an additional term due to the stretching by the shear.

Kullenberg[106] determined the horizontal variances σ^2_x and σ^2_y as well as K_z experimentally by means of dye tracer observations. The mean current shear and oscillations were determined from current measurements carried out in conjunction with the tracer experiments. Thus it was possible to compare observed and predicted variances for time periods up to several days. The agreement between observations and predictions was fairly good (Figure 21), suggesting that the vertical shear effect could explain a large part of the observed horizontal dispersion.

It is interesting to note that to some extent the shear diffusion model and the similarity models conform as regards the predictions. Considering the unbounded case for relatively large diffusion times the horizontal variance is

$$\sigma_x^2 \propto K_z \left(\frac{dU_o}{dz}\right)^2 t^3 \tag{99}$$

The effective horizontal mixing coefficient is then, assuming that K_z and a_o are constant,

$$K_x \propto \frac{d\sigma_x^2}{dt} \propto K_z \left(\frac{dU_o}{dz}\right)^2 t^2 \tag{100}$$

Elimination of time t yields

$$K_x \propto \sigma_x^{4/3} \left[K_x \left(\frac{dU_o}{dz}\right)^2 \right]^{1/3} \tag{101}$$

It is interesting to note that the scale dependence conforms with Richardson's scale dependence law.

It is important to distinguish between the bounded case, treated by Bowden[252] when the variance is inversely proportional to K_z and the unbounded case. Evans[256] gave an exact solution for a model of shear diffusion between two well-mixed layers. He demonstrated that great care is required when using the approximations for the bounded or unbounded case, respectively, since there is a large time interval when the exact solution only is applicable. For $K_z t/h^2 \leqslant 0.1$ the unbounded case can be considered as a good approximation, and the bounded case is an acceptable approximation when $K_z t/h^2 \geqslant 1.7$. According to these criteria the unbounded case should be applicable for Kullenberg's[106] experiments.

XI. PARAMETERIZATION OF THE MIXING

Here a summary will be given of attempts to relate the mixing to the environmental conditions. It is pertinent to begin with considering the mixing in the surface layer down to the primary pycnocline. Several one-dimensional models have been developed for the wind-mixed layer, the generation and erosion of the seasonal thermocline.[257-259]

The wind is often the main driving factor and the vertical mixing can be related to the wind. Kullenberg[106] proposed the relationship

$$K_z = \left(Rf \cdot c_d \cdot \frac{\rho_a}{\rho_o} \right) \frac{\overline{W_{10}^2}}{\overline{N^2}} \left| \frac{dq}{dz} \right|, \quad W_{10} > 5 \text{ m s}^{-1} \tag{102}$$

where dq/dz is the gradient of the current vector. The numerical coefficient is about 10^{-7}. The relationship is shown in Figure 22. The vertical mixing coefficient was determined from subsurface dye experiments carried out during widely different conditions in various areas. The depth of the dye varied between about 5 and 40 m.

Earlier we have seen that the momentum transfer may be given in the form, for the surface layer.

$$A_z = \frac{1}{f}\left(\frac{\rho_a c_d}{\rho_0 k_W}\right)^2 W_{10}^2 \tag{103}$$

The ratio K_z/A_z is found to be

$$\frac{K_z}{A_z} = \frac{Rf \cdot k_W^2}{c_d}\frac{\rho_0}{\rho_a}\frac{f}{N} Ri^{-\frac{1}{2}} \tag{104}$$

By means of the relation

$$\frac{\partial v}{\partial z} = \frac{g}{f \cdot \bar{\rho}}\frac{\partial \rho}{\partial y} \tag{105}$$

the slope i of a density surface can be given with the form

$$i = \frac{f}{N}\cdot Ri^{-\frac{1}{2}} \tag{106}$$

which implies that $K_z/A_z \propto i$. This shows that in general the momentum transfer is much more efficient than the diffusion, or transfer of substances. It is in agreement with the Monin-Obukhov theory and the relationship

$$\frac{K_z}{A_z} = \frac{Rf_c}{Ri} \tag{107}$$

Experimental evidence from the field and the laboratory seems to be consistent with such a relationship for $Ri > 0.4$.[260]

Combining Equations 102 and 107 and using the definition of the Richardson number Ri, it is found that

$$A_z = \frac{Ri}{Rf_c}\cdot K_z = \frac{1}{Rf_c}\left[\frac{N^2}{\left(\dfrac{dq}{dz}\right)^2}\right]\cdot\frac{Rf\cdot c_d\cdot\rho_a}{\rho_0}\frac{W_{10}^2}{N^2}\left|\frac{dq}{dz}\right| \tag{108}$$

$$A_z = \frac{Rf\cdot c_d\cdot\rho_a}{Rf_c\cdot\rho_0}\cdot\frac{W_{10}^2}{\left|\dfrac{dq}{dz}\right|}$$

This can be rewritten in the form

$$A_z\cdot\rho_0\left|\frac{dq}{dz}\right| = \left|\tau_0\right| = \frac{Rf\cdot c_d\cdot\rho_a}{Rf_c}\cdot W_{10}^2 \tag{109}$$

in agreement with Equation 61 and showing that the results are consistent.

The parameterization of K_z as driven by the wind can only apply when the wind is the primary source of energy for the mixing in stable conditions. Thorpe[179] used observations of the overturning, or Ozmidov, scale in a lake to parameterize the vertical mixing. The rate of energy dissipation per unit mass is then

$$\epsilon = const. \cdot \ell^2 N^3$$

where ℓ is the overturning scale. Ozmidov[167] argued that ℓ is the limiting scale for the buoyancy influence on the motion and that for smaller scales the turbulence is isotropic. This leads to the expression

$$K_{z\,max} = \text{const.} \cdot \epsilon^{1/3} \cdot \varrho^{4/3} = \text{const.} \cdot N \cdot \varrho^2 \qquad (110)$$

which Thorpe[179] used to calculate K_z from his observations.

In many areas on the shelf seas the tidal mixing will be equal to or even larger than the wind-induced mixing. Pingree et al.[239] showed that a wind speed of about 13 ms^{-1} would be equal to a tidal current velocity of maximum 1 knot during a tidal cycle.

The mixing across a layer of increasing density (pycnocline layer) has with good results been studied in laboratory experiments.[134, 257] In conditions of a wind-mixed layer reaching down to the top of a fairly sharp thermocline, fluid from below the thermocline becomes entrained into the upper layer. Thereby the thermocline is eroded and gradually deepened. The rate of deepening defines the so-called entrainment velocity w_e which dimensionally can be expressed as

$$w_e = u_* \cdot f(Ri_*) \qquad (111)$$

where Ri_* is an overall or bulk Richardson number defined by

$$Ri_* = \frac{g\,\Delta\rho\,h}{\rho_0\,u_*^2} \qquad (112)$$

Here $\Delta\varrho$ is the density difference between the layers and h is the thickness of the upper layer. Over a limited range of Richardson numbers experimental evidence shows[260] (Figure 23).

$$w_e = c\,u_*\,Ri_*^{-1}, \quad 50 \leqslant Ri_* \leqslant 5 \cdot 10^3 \qquad (113)$$

The relation between K_z and w_e can be found from the expressions for the vertical flux Q of mass per unit area

$$Q = -K_z\,\frac{d\bar{\rho}}{dz} = \frac{K_z\,N^2}{g/\rho_0} = w_e\,\Delta\rho \qquad (114)$$

$$K_z = g\,\frac{\Delta\rho}{\rho_0}\,w_e\,N^{-2} \qquad (115)$$

Introducing the expression for w_e and the definition of Ri_*, yields

$$K_z = c\,\frac{u_*^3/h}{N^2} \qquad (116)$$

The rate of energy dissipation per unit mass can be expressed as

$$\epsilon = \frac{u_*^3}{\kappa_0 h} \qquad (117)$$

leading to

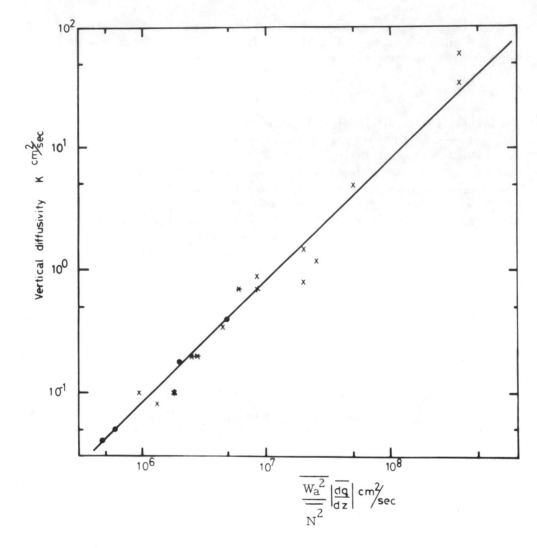

FIGURE 22.   Vertical exchange (diffusion) coefficient[106]

$$K_z \text{ versus } \overline{W_a^2} \cdot \overline{N^{-2}} \cdot \left| \frac{d\overline{q}}{dz} \right|, \ W_a > 5m \cdot s^{-1}$$

(From Kullenberg, G., An Experimental and Theoretical Investigation of the Turbulent Diffusion in the Upper Layer of the Sea, Vol. 25, Institute for Physical Oceanography, University of Copenhagen, 1974.)

$$K_z = c \kappa_o \epsilon N^{-2} \tag{118}$$

With $c \ \kappa_o \sim 1$ and $\kappa_o \sim 0.4$ we find $c \sim 2.5$ which is in good agreement with the laboratory experiments by Kato and Phillips.[261] On the other hand, the values of Osborn (Equation 80) would imply $c \ \kappa_o \sim 0.2$ and $c \sim 0.5$ with $\kappa_o = 0.4$.

Garrett and Munk[43,44] developed a formalism for mixing in the interior of the ocean due to breaking of internal waves through shear instabilities. They used their universal internal wave spectrum and assumed further that the shear layers grow at a rate required to maintain a critical Richardson number of 0.4 ± 0.1 in the layer.[262] The corresponding mixing coefficient was given as

$$K_z = \frac{1}{12} \int_0^\infty h^3 \ n \ (h) \ dh \tag{119}$$

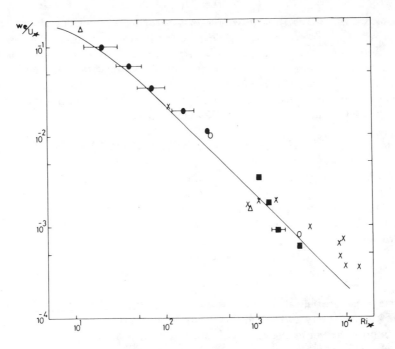

FIGURE 23. Ratio of entrainment velocity to friction velocity w_e/u^* vs. overall Richardson number Ri*, from experiments in the sea, in lakes, and in the laboratory. (With permission from Kullenberg, G., *Estuarine Coastal Mar. Sci.*, 5, 329, 1977. Copyright by Academic Press, Inc. Ltd.)

Here n(h) dh is the expected number of breaking events between h ± ½ dh, h being the depth interval under consideration. On the basis of expected values of various parameters Garrett and Munk[43,44] estimated $K_z \simeq 5 \cdot 10^{-6}$ m²s⁻¹. They also concluded that the mixing would decrease markedly with depth.

The mixing in the bottom boundary layer may in principle be parameterized in the same way as in the surface layer. The important layers are the Ekman boundary layer h_E and the logarithmic layer h_{ln}

$$h_E = \frac{\kappa_o u_*}{f}, \quad h_{ln} = \frac{2 u_*^2}{f U_g} \quad \text{or} \quad h_{ln} = 0.1 h_E \qquad (120)$$

In the logarithmic layer the momentum transfer can in neutral conditions be given as

$$A_z = -\kappa_o u_* z \qquad (121)$$

or it may be related to the layer thickness as

$$A_z = \frac{u_* H_E}{c_1} = \frac{\kappa_o u_*^2}{c_1 f} \qquad (122)$$

where $c_1 \sim 15-25$. The diffusion coefficient may in near-neutral conditions be considered as equal to the momentum transfer coefficient. Alternatively we have

$$K_z = Rf_c \epsilon N^{-2} = Rf_c u_*^2 \left(\frac{dU}{dz}\right) N^{-2} \qquad (123)$$

This expression appears to agree with the observations of Sarmiento et al.[220]

The various possibilities of parameterization of the mixing in terms of environmental conditions will usually make it possible to estimate the mixing conditions once the range of environmental variability in an area is known. However, very often the basic difficulties lie in obtaining reliable and representative observations of the environmental conditions. In particular information on the extreme conditions can be difficult to obtain. Therefore it is in many cases necessary to turn to the development of various predictive models, as discussed in the following chapter.

XII. CONCLUDING REMARKS

A large number of processes, occurring over a wide range of scales in time and space, effect the dispersion of pollutants in the ocean. We have discussed processes in boundary layers coupled to the surface and the solid boundaries, processes occurring along oceanic fronts, in the interior of the ocean related to mesoscale eddies, spreading along isopycnic surfaces, small-scale instabilities, and molecular effects. In many cases we are as yet only aware of possible mechanisms but cannot isolate, quantify, or parameterize these. Hence we cannot determine the relative significance of the various processes.

We have considerable information on the distribution of natural properties in the ocean, the heterogeneity and variability being basic characteristics. However, in many cases we cannot as yet ascertain which are the basic dynamical processes generating these distributions.

Pollutants are mostly introduced at the boundaries and from the atmosphere, as well as from the land directly. Basic questions in relation to the distribution of pollutants therefore concern the transfer in the boundary layers, between them and from them into the interior of the ocean. We may get a picture of the transfer phases through the following conceptual model. The physics of the frictional boundary layers at the surface and the bottom are the same. A pollutant introduced in one of these layers will tend to become vertically mixed over the layer with subsequent horizontal spreading in the layer. The transfer from the boundary layers into the interior is weak, except in certain regions, and may be coupled to seasonal variations or episodic events. There are indications that the vertical mixing is relatively more efficient along the continental boundaries than in the interior. From the boundary layers along the slope, the spreading into the interior mainly occurs along isopycnic surfaces. Efficient transfer from the surface layer into the interior may also occur at oceanic fronts.

This is, however, too simplified a picture with several omissions. In some limited areas of deep water formation very efficient transfer can occur from the surface layer to the deep and bottom water masses. High-speed western boundary currents transfer material towards high latitudes where the vertical transfer is relatively efficient. The high-energy, meso-scale motion can probably also generate vertical exchange between the surface and bottom boundary layers, and the interior. It is also clear that some finite cross-isopycnal mixing as well as vertical motion occur in the interior. Episodic mixing events can generate well-mixed patches, which will spread along isopycnic surfaces as intrusions. However, intrusions and active microstructure processes are most commonly observed in relation to frontal zones. In the case of mixing along isopycnic surfaces the rate of vertical transfer will depend upon the slope of the surfaces, which is generally most pronounced at high latitudes.

The atmosphere is an important source of pollutants. Part of this input can get transferred fairly directly to the interior via deep and bottom water formation areas and subsequent spreading with these waters. In this way many substances can reach most regions of the interior of the oceans. Direct transfer to the bottom may also occur for some substances through particle settling.

Outstanding features of the ocean are the stable stratification, implying that wave motion is very important, the meso-scale horizontal inhomogenieties, and the small- and microscale structures in the vertical distributions occurring in most regions. Pollutants will to a certain extent become distributed in a similar way. This is confirmed experimentally for the top-most kilometer, and it is not unlikely that pollutants may become distributed in such a fashion also in deep and bottom waters. Layers can be very persistent with considerably higher concentrations than the surrounding water.

In many regions the mixing is very weak occurring mainly on an intermittent basis. Relatively high concentrations of pollutants may therefore be encountered at intervals in different layers of the water column.

The motion in the sea is composed of several components, apart from the large-scale oceanic circulation gyres. It is important to separate between wave motion, which does not give rise to mixing except when the waves break, advection, meso-scale eddy motion and tidal motion stirring the ocean, and three-dimensional turbulent motion generating mixing and cross-isopycnal exchange. The effect of the fluctuating motion on the distribution of properties if often quantified by a semi-empirical parameterization. Generalizations of such formulations are often questionable. As yet, the use of turbulent diffusion coefficients (or eddy diffusion coefficients) is the most common way to express the mixing.

It is to a large extent the research of the last decades which has revealed several of the features discussed here. Almost continuous development of sophisticated, accurate instruments may make it possible to distinguish various processes in the ocean and test theoretical ideas. This may eventually make it possible to predict reliably the dispersion of pollutants.

It is of course important also to note that pollutants have very different properties, chemically, biologically, and physically, which all should be taken into account when considering their dispersion in the ocean. Some of these aspects are elaborated in later chapters in this book. In relation to time scales of oceanic mixing and transport, the different persistency of pollutants should be borne in mind, ranging from infinity for metals, very long for substances as DDT and PCBs, to short-lived for bacteria and virus.

NOTATIONS*

A_{1z}, A_z	Vertical turbulent exchange coefficient for momentum
A_x, A_y, A_h	Horizontal turbulent exchange coefficient for momentum
B	Buoyancy flux
b_*	Buoyancy
C	Concentration of diffusing substance
Co	Cox number
c_b, c_d	Bottom and surface drag (friction) coefficients, respectively
E_w	Wind energy per unit surface
$f = 2\omega_o \sin\phi$	Coriolis parameter
Fr	Densimetric Fronde number
g	Acceleration of gravity
$g' = g \cdot \Delta\varrho/\varrho$	Buoyancy (or reduced gravity)
H	Water depth
h	Vertical scale or boundary layer thickness
i	Slope of isopycnals or isopycnic surfaces
k	Wave number
k_w	Wind factor
K_h, K_z	Horizontal and vertical turbulent diffusion coefficient, respectively
K_{zT}, K_{zS}	Vertical turbulent diffusion coefficient for heat and salt, respectively
L	Monin-Obukhov length scale
l	Turbulence length scale
l_h	Horizontal length scale
$N = [- g/\varrho$ $d\varrho/dz]^{1/2}$	Brunt-Väisälä frequency
P	Horizontal diffusion velocity
p, p'	Pressure and pressure fluctuations, respectively
q	Horizontal current vector
R_d	Rossby radius of deformation
Rf, Rf_c	Flux and critical flux Richardson number, respectively
Rf^T	Bulk flux Richardson number
Ri, Ri_c, Ri_*	Richardson, critical Richardson, and overall or bulk Richardson number, respectively
r	Equivalent radius
S, s	Salinity and salinity fluctuations, respectively
t	Time
T, T'	Temperature and temperature fluctuations, respectively
U, V, W	Mean velocities in x, y, z directions, respectively
u, v, w	Velocities in x, y, z directions, respectively
u', v', w'	Fluctuating velocities in x, y, z directions, respectively
U_o	Steady current component in center of diffusing layer
u_o, v_o	Oscillating current components in center of diffusing layer
u_*	Friction velocity
w_e	Entrainment velocity
W_s, W_{10}	Wind velocity at surface and 10-m level, respectively
x, y, z,	Orthogonal coordinates, positive east, north, and upwards, respectively, with z = 0 at sea surface
α	Volume expansion coefficient for heat

---

* Some notations used only once are explained in the text; bar implies time averaging.

β_s	Volume expansion coefficient for salt
ε	Rate of energy dissipation per unit mass
ε_T	Rate of dissipation of temperature fluctuations
κ_T, κ_S	Molecular diffusion coefficients for heat and salt, respectively
κ_o	von Karmen constant
μ, ν	Dynamic and kinematic viscosity, respectively
ϱ_o, ϱ_a	Density of water and air, respectively
ϱ_o	Mean density of water
$\Delta\varrho$	Density difference across interface or pycnocline layer
ω	Angular velocity of oscillating current component
ω_o	Angular velocity of the rotation of the earth latitude
ϕ	Latitude
λ	Wavelength
ζ	Surface elevation
σ_{rc}^2	Radial variance of spherically symmetrical concentration distribution
σ_x^2, σ_y^2	Variance of concentration distribution in x and y directions, respectively
σ_{rh}^2	Horizontal and radial variance of concentration distribution

REFERENCES

1. **Launder, B. E. and Spalding, D. B.**, *Mathematical Models of Turbulence,* Academic Press, New York, 1972.
2. **Tennekes, H. and Lumley, J. L.,** *A First Course in Turbulence,* MIT Press, Cambridge, Mass., 1972.
3. **Eckart, C.,** An analysis of the stirring and mixing processes in incompressible fluids, *J. Mar. Res.,* 7, 265, 1948.
4. **Newmann, G. and Pierson, W. J.,** *Principles of Physical Oceanography,* Prentice-Hall, Englewood Cliffs, N.J., 1966.
4a. **Moore, H. B.,** Aspects of stress in the tropical marine environment, *Adv. Mar. Biol.,* 10, 217, 1972.
5. **Weidemann, H., Ed.,** The ICES diffusion experiment Rheno 1965, *Rapp. P. V. Reun. Cons. Int. Explor. Mer,* 163, 1, 1973.
6. **Thompson, R.,** Topographic Rossby waves at a site north of the Gulf Stream, *Deep Sea Res.,* 18, 1, 1971.
7. **Webster, F.,** The effect of meanders on the kinetic energy balance of the Gulf Stream, *Tellus,* 13, 392, 1961.
8. **Swallow, J. C.,** The Aries current measurements in the western North Atlantic, *Phil. Trans. R. Soc. London A,* 270, 451, 1971.
9. **Koshlyakov, M. N. and Grachev, Y. M.,** Meso-scale currents at a hydrographical polygon in the tropical Atlantic, *Deep Sea Res.,* 20, 507, 1973.
10. **Gould, W. J., Schmitz, W. J., Jr., and Wunsch, C.,** Preliminary field results for a Mid-Ocean Dynamics Experiment (MODE - O), *Deep Sea Res.,* 21, 911, 1974.
11. **Freeland, H. J., Rhines, P. B., and Rossby, H. T.,** Statistical observations of the trajectories of neutrally buoyant floats in the North Atlantic, *J. Mar. Res.,* 33, 383, 1975.
12. **The MODE Group,** The Mid-Ocean Dynamics Experiment, *Deep Sea Res.,* 25, 859, 1978.
13. **Wyrtki, K., Magaard, L., and Hager, J.,** Eddy energy in the oceans, *J. Geophys. Res.,* 81, 2641, 1976.
14. **Schmitz, W. J., Jr.,** Eddy kinetic energy in the deep western North Atlantic, *J. Geophys. Res.,* 81, 4981, 1976.
15. **Schmitz, W. J., Jr.,** On the deep general circulation of the western North Atlantic, *J. Mar. Res.,* 35, 21, 1977.
16. **Schmitz, W. J., Jr.,** Observations of the vertical distribution of low frequency kinetic energy in the western North Atlantic, *J. Mar. Res.,* 36, 295, 1978.
17. **Luyten, J. R.,** Scales of motion in the deep Gulf Stream and across the Continental Rise, *J. Mar. Res.,* 35, 49, 1977.

18. Phillips, O. M., *The Dynamics of the Upper Ocean,* 1st ed., Cambridge University Press, New York, 1966.
19. Crease, J., Velocity measurement in the deep of the western North Atlantic, summary, *J. Geophys. Res.,* 67, 3173, 1962.
20. Swallow, J. C., A deep eddy off Cape St. Vincent, *Deep Sea Res.,* 16 (Suppl.), 285, 1969.
21. Fuglister, F. C. and Worthington, L. V., Some results of a multiple ship survey of the Gulf Stream, *Tellus,* 3, 1, 1951.
22. Fuglister, F. C., Gulf Stream '60, *Prog. Oceanogr.,* 1, 265, 1963.
23. Fuglister, F. C., Cyclonic rings formed by the Gulf Stream, 1965—1966, in *Studies in Physical Oceanography,* Vol. 1, Gordon, A. L., Ed., Gordon and Breach, New York, 1972, 137.
24. Parker, C. E., Gulf Stream rings in the Sargasso Sea, *Deep Sea Res.,* 18, 981, 1971.
25. Fuglister, F. C., A cylconic ring formed by the Gulf Stream, 1967, in *A Voyage of Discovery,* Angel M., Ed., Pergamon Press, New York, 1977, 177.
26. Richardson, P. L., Cheney, R. E., and Mantini, L. A., Tracking a Gulf Stream ring with a free drifting surface buoy, *J. Phys. Oceanogr.,* 7, 580, 1977.
27. Saunders, P. M., Anticyclonic eddies formed from shoreward meanders of the Gulf Stream, *Deep Sea Res.,* 18, 1207, 1971.
27a. Fischer, H. B., Mixing processes on the Atlantic continental shelf Cape Cod to Cape Hatteras, *Limnol. Oceanogr.,* 25(1), II4, 1980.
27b. Smith, P. C., Low-frequency fluxes of momentum, heat, salt and nutrients at the edge of the Scotian shelf, *J. Geophys., Res.,* 83(C8), 4079, 1978.
28. Wiebe, P. H., Hulburt, E. M., Carpenter, E. J., Jahn, A. E., Knapp, G. P., Boyd, S. H., Ortner, P. B., and Cox, J. L., Gulf Stream cold core rings: large-scale interaction sites for open ocean plankton communities, *Deep Sea Res.,* 23, 695, 1976.
29. Wiebe, P. H., and Boyd, S. H., Limits of Nematoscelis megalops in the north-western Atlantic in relation to Gulf Stream cold core rings. I. Horizontal and vertical distributions, *J. Mar., Res.,* 36, 119, 1978.
30. Joyce, T. M. and Patterson, S. L., Cyclonic ring formation at the polar front in Drake Passage, *Nature (London),* 265, 131, 1977.
31. Cutchin, D. L. and Smith, R. L., Continental shelf waves: low-frequency variations in sea level and currents over the Oregon continental shelf, *J. Phys. Oceanogr.,* 3, 73, 1973.
32. Thompson, R. and Luyten, J. R., Evidence for bottom-trapped topographic Rossby waves from single moorings, *Deep Sea Res.,* 23, 629, 1976.
33. Smith, P. C., Baroclinic instability in the Denmark Strait overflow, *J. Phys. Oceanogr.,* 6, 355, 1976.
34. Thorpe, S. A., Variability of the Mediterranean undercurrent in the Gulf of Cadiz, *Deep Sea Res.,* 23, 711, 1976.
35. Niiler, P. P. and Richardson, W. S., Seasonal variability of the Florida Current, *J. Mar. Res.,* 31, 144, 1973.
36. Orlanski, I., The influence of bottom topography on the stability of jets in a baroclinic fluid, *J. Atmos. Sci.,* 26, 1216, 1969.
37. Schott, F. and Düing, W., Continental Shelf waves in the Florida Straits, *J. Phys. Oceanogr.,* 6, 451, 1976.
38. Voorhis, A. D., Schroeder, E. H., and Leetmaa, A., The influence of deep mesoscale eddies on sea surface temperature in the North Atlantic subtropical convergence, *J. Phys. Oceanogr.* 6, 963, 1976.
39. Pollard, R. T. and Millard, R. C., Comparison between observed and simulated wind-generated inertial oscillations, *Deep Sea Res.,* 17, 813, 1970.
40. Gustafsson, T. and Kullenberg, B., Untersuchungen von Trägheitsströmungen in der Ostsee, *Sven. Hydrogr. Biol. Komm. Skr. Ny Ser. Hydrogr.,* 13, 28 pp., 1936.
41. Garrett, C., Mixing in the ocean interior, *J. Dynam. Atmos. Oceans,* 3(2—4), 1979.
42. Phillips, O. M., *Dynamics of the Upper Ocean,* 2nd ed., Cambridge University Press, London, 1977.
43. Garrett, C. and Munk, W., Oceanic mixing by breaking internal waves, *Deep Sea Res.,* 19, 823, 1972.
44. Garrett, C. and Munk, W., Space-time scales of internal waves: a progress report, *J. Geophys. Res.,* 80, 291, 1975.
45. Bowden, K. F., Measurements of turbulence near the sea bed in a tidal current, *J. Geophys. Res.,* 67, 3181, 1962.
46. Bowden, K. F. and Howe, M. R., Observations of turbulence in a tidal current, *J. Fluid Mech.,* 17, 271, 1963.
47. Ellison, T. H., Turbulent transport of heat and momentum from an infinite rough plane, *J. Fluid Mech.,* 2, 456, 1957.
48. Stewart, R. W., The problem of diffusion in stratified fluid, *Adv. Geophys.,* 6, 303, 1959.
49. Wimbush, M. and Munk, W., The benthic boundary layer, in *The Sea,* Vol. 4, Maxwell, A., Ed., Interscience, New York, 1971, 731.

50. **Weatherly, G. L.,** A study of the bottom boundary layer of the Florida current, *J. Phys. Oceanogr.,* 2, 54, 1972.

51. **Grant, H. L., Stewart, R. W., and Moilliet, A.,** Turbulence spectra from a tidal channel, *J. Fluid Mech.,* 12, 241, 1962.

52. **Grant, H. L., Moilliet, A., and Vogel, W. M.,** Some observations of the occurrence of turbulence in and above the thermocline, *J. Fluid Mech.,* 34, 443, 1968.

53. **Batchelor, G. K.,** Small-scale variation of convected quantities like temperature in turbulent fluid, I, *J. Fluid Mech.,* 5, 113, 1959.

54. **Nasmyth, P.,** Oceanic Turbulence, Ph.D. thesis, University of British Columbia, Vancouver, 1970.

55. **Taylor, G. I.,** Diffusion by continuous movement, *Proc. London, Math. Soc. Ser. 2,* 20, 196, 1921.

56. **Woods, J. D.,** Wave-induced shear instability in the summer thermocline, *J. Fluid Mech.,* 32, 791, 1968.

57. **Worthington, L. V.,** The Norwegian Sea as a Mediterranean basin, *Deep Sea Res.,* 17, 77, 1970.

57a. **Needler, G. T.,** Comments on high latitudes processes for oceanclimate modelling, *J. Dynam. Atmos. Oceans,* 3, 2, 1979.

58. **Killworth, P. D.,** Mixing during MEDOC, *Prog. Oceanogr.* 7(2), 1, 1976.

58a. **Iselin, C. O'D.,** The influence of vertical and lateral turbulence on the characteristics of the waters at mid-depths, *Am., Geophys. Union Trans. Pt. 3,* 414, 1939.

59. **Munk, W.,** Abyssal recipes, *Deep Sea Res.,* 13, 707, 1966.

60. **Mooers, C. N. K., Collins, C. A., and Smith, R. L.,** The dynamic structure of the frontal zone in the coastal upwelling region off Oregon, *J. Phys. Oceanogr.,* 6, 3, 1976.

61. **Walin, G.,** On the hydrographical response to transient meteorological disturbance, *Tellus,* 24, 169, 1972.

62. **Shaffer, G.,** Calculations of Cross-Isohaline Salt Exchange in a Coastal Region of the Baltic, Rept. No. 24, Oceanographic Institute, University of Gothenburg, Gothenburg, Sweden, 1977.

63. **Joyce, T. M., Zenk, W., and Toole, J. M.,** The anatomy of the Antarctic Polar Front in the Drake Passage, *J. Geophys. Res.,* 83, 6093, 1978.

64. **Kullenberg, G.,** Light-scattering observations in frontal zones, *J. Geophys. Res.,* 83, 4683, 1978.

65. **Roden, G. I.,** Oceanic subarctic fronts of the central Pacific. Structure of and response to atmospheric forcing, *J. Phys. Oceanogr.,* 7, 761, 1977.

66. **Sverdrup, H. U., Johnson, M. W., and Fleming, R. H.,** *The Oceans: Their Physics, Chemistry and General Biology,* Prentice-Hall, New York, 1942.

67. **Knauss, J. A.,** Measurements of the Cromwell Current, *Deep Sea Res.,* 6, 265, 1960.

68. **Pingree, R. D.,** Mixing and stabilization of phytoplankton distributions on the Northwest European Continental Shelf, in *Spatial Pattern in Plankton Communities,* Steele, J. H., Ed., Plenum Press, New York, 1978.

69. **Bowman, M. J. and Esaias, W. E.,** *Oceanic Fronts in Coastal Processes,* Springer-Verlag, Berlin, 1978.

70. **Woods, J. D., Wiley, R. L., and Briscoe, M. G.,** Vertical circulation at fronts in the upper ocean, in *A Voyage of Discovery,* Angel, M., Ed., 253, 1977.

71. **Munk, W. H.,** On the wind-driven ocean circulation *J. Meteorol.,* 7, 79, 1950.

72. **Stommel, H.,** *The Gulf Stream,* University of California Press, Berkeley, 1965.

73. **Okubo, A.,** Horizontal and vertical mixing in the sea, in *The Impingement of Man on the Oceans,* Hood, D. W., Ed., Interscience, New York, 1971, 89.

74. **Defant, F.,** *Physical Oceanography,* Vol. I, Pergamon Press, Oxford, 1961.

75. **Veronis, G.,** Model of world ocean circulation. III. Thermally and wind driven, *J. Mar. Res.,* 36, 1, 1978.

76. **Kuo, H. H. and Veronis, G.,** The use of oxygen as a test for an abyssal circulation model, *Deep Sea Res.,* 20, 871, 1973.

77. **Veronis, G.,** Use of tracers in circulation studies, in *The Sea,* Vol. 6, Goldberg, E. D., McCave, I. N., O'Brien, J.J., Steele, J. H., Eds., Interscience, New York, 1977, 169.

78. **Fiadeiro, M. E. and Craig, H.,** Three-dimensional modeling of tracers in the deep Pacific Ocean. I. Salinity and oxygen, *J. Mar. Res.,* 36, 323, 1978.

79. **Stommel, H. and Arons, A. B.,** On the abyssal circulation of the world oceans. II. An idealised model of the circulation pattern and amplitude in oceanic basins, *Deep Sea Res.,* 6, 217, 1960.

80. **Harvey G. R. and Steinhauer, W. G.,** Transport pathways of polychlorinated biphenyls in Atlantic water, *J. Mar. Res.,* 34, 561, 1976.

81. **Rooth, C. G., and Östlund, H. G.,** Penetration of tritium into the Atlantic thermocline, *Deep Sea Res.,* 19, 481, 1972.

82. **Östlund, H. G., Dorsey, H. G., and Rooth, C. G.,** GEOSECS North Atlantic radio carbon and tritium results, *Earth Planetary Sci. Lett.,* 23, 69, 1974.

83. **Kullenberg, G.,** Observations of the mixing in the Baltic thermo- and halocline layers, *Tellus,* 29, 572, 1977.

84. **Allain, C.,** Topographic dynamique et courants généreux dans le bassin occidental de la Méditerranée, *Rev. Trav. Inst. Pechesmarit,* 24, 121, 1960.
85. **Gerges, M. A.,** Preliminary results of a numerical model of circulation using density field in the eastern Mediterranean, *Acta Adriatica,* 18, 165, 1976.
86. **Lacombe, H. and Tchernia, P.,** Characteres hydrologiques et circulation des eaux en Mediterranee, in *The Mediterranean Sea,* Stabley, D. J., Ed., Dowden, Hutchinson and Ross, Inc., 1972, 25.
87. **Stommel, H. and Fedorov, K. N.,** Small scale structure in temperature and salinity near Timor and Mindanao, *Tellus,* 14, 306, 1967.
88. **Cooper, J. and Stommel, H.,** Regularly spaced steps in the main thermocline near Bermuda, *J. Geophys. Res.,* 73, 5849, 1968.
89. **Pingree, R. D.,** Small-scale structure of temperature and salinity near station Cavall, *Deep Sea Res.,* 16, 275, 1969.
90. **Cox, G. S., Nagata, Y., and Osborn, T.,** Oceanic fine structure and internal waves, *Bull. Jpn. Soc. Fish. Oceanogr.,* Special No. 67, 1969.
91. **Woods, J. D. and Wiley, R. L.,** Billow turbulence and ocean microstructure, *Deep Sea Res.,* 19, 87, 1972.
92. **Osborn, T. R. and Cox, C. S.,** Oceanic fine structure, *Geophys. Fluid Dyn.,* 3, 321, 1972.
93. **Howe, M. R. and Tait, R. J.,** The role of temperature inversions in the mixing processes in the deep ocean, *Deep Sea Res.,* 19, 781, 1972.
94. **Molcard, R. and Williams, A. J.,** Deep stepped structure in the Tyrrhenian Sea, *Memoires Societe R. Sci. Liege, 6e Ser.,* 7, 191, 1975.
95. **Gregg, M. C.,** Temperature and salinity microstructure in the Pacific Equatorial Undercurrent, *J. Geophys. Res.,* 81, 1180, 1976.
96. **Gregg, M. C.,** Microstructure and intrusion in the California Current, *J. Phys. Oceanogr.,* 5, 253, 1975.
97. **Gregg, M. C.,** A comparison of finestructure spectra from the main thermocline, *J. Phys. Oceanogr.,* 7, 33, 1977.
98. **Gargett, A. E.,** An investigation of the occurrence of oceanic turbulence with respect to finestructure, *J. Phys. Oceanogr.,* 6, 139, 1976.
99. **Fedorov, K. N.,** *The Thermohaline Finestructure of the Ocean,* Pergamon Press, New York, 1978.
100. **Simpson, J. and Woods, J. D.,** The temperature microstructure in a freshwater thermocline, *Nature (London),* 226, 832, 1970.
101. **Hoare, R. A.,** Thermohaline convection in Lake Vanda, Antarctica, *J. Geophys. Res.,* 73, 607, 1968.
102. **Foldvik, A., Thomsen H., and Westerberg, H.,** Temperature Microstructure in Fjords, Rept. No. 33, Geophysical Institute, University of Bergen, Bergen, Norway, 1973.
103. **Sanford, T. B.,** Observations of the vertical structure of internal waves, *J. Geophys. Res.,* 80, 3861, 1975.
104. **Osborn, T. R.,** Measurements of energy dissipation adjacent to an island, *J. Geophys. Res.,* 83, 2939, 1978.
105. **Kullenberg, G.,** Apparent horizontal diffusion in a stratified vertical shear flow, *Tellus,* 24, 17, 1972.
106. **Kullenberg, G.,** *An Experimental and Theoretical Investigation of the Turbulent Diffusion in the Upper Layer of the Sea,* Vol. 25, Institute of Physical Oceanography, University of Copenhagen, 1974.
107. **Kullenberg, G.,** Investigation of small-scale vertical mixing in relation to the temperature structure in stably stratified waters, *Adv. Geophys.,* 18A, 339, 1974.
108. **Hansen, J.,** *Tracer Engineering in Coastal Pollution Control,* IAEA SM-142/37, International Atomic Energy Agency, Vienna, 1963.
109. **Abraham, G.,** Jet Diffusion in Stagnant Ambient Fluid, Publ. no. 29, Delft Hydraulics Laboratory, The Hague, 1963.
110. **Gibbs, R. J.,** Ed., *Suspended Solids in Water,* Plenum Press, New York, 1974.
111. **Jerlov, N. G.,** *Marine Optics,* Elsevier, Amsterdam, 1976.
112. **Kitchen, J. C., Zaneweld, J. R. V., and Pak, H.,** The vertical structure and size distribution of suspended particles off Oregon during the upwelling season, *Deep Sea Res.,* 25, 453, 1978.
113. **Jerlov, N. G.,** Distribution of suspended material in the Adriatic Sea, *Arch. Oceanogr. Limnol.,* 11, 227, 1958.
114. **Pak, H., Beardsley, G. F., Jr., and Pak, P. K.,** The Columbia River as a source of marine light scattering particles, *J. Geophys. Res.,* 75, 4570, 1970.
115. **Gibbs, R. J.,** The suspended material of the Amazon shelf and tropical Atlantic Ocean, in *Suspended Solids in Water,* Gibbs, R. J., Ed., Plenum Press, New York, 1974, 203.
116. **Rex, R. W. and Goldberg, E. D.,** Quartz contents of pelagic sediments of the Pacific Ocean, *Tellus,* 10, 153, 1958.

117. Biscaye, P. E. and Eittreim, S. L., Variations in benthic boundary layer phenomena: neopheloid layer in the North American basin, in *Suspended Solids in Water*, Gibbs, R. J., Ed., Plenum Press, New York, 1974, 227.

118. Kullenberg, G., Light scattering observations in the northwest African upwelling region, *Deep Sea Res.*, 25, 525, 1978.

119. GESAMP, Scientific criteria for the selection of sites for dumping of wastes into the sea, *GESAMP Reports and Studies*, Vol. III, 1975.

120. Dietrich, G., Kalle, K., Krauss, W., and Siedler, G., *Allgemeine Merreskunde*, 3, Auflage, Gebrüder Borntraeger, Berlin, 1975.

121. Stewart, R. W., in Turbulence in the Ocean, Report on the Symposium 1968, Rept. no. 34, University of British Columbia, Vancouver, 1970.

122. Gregg, M. C. and Cox, C. S., The vertical microstructure of temperature and salinity, *Deep Sea Res.*, 19, 355, 1972.

123. Hayes, S. P., Joyce, T. M., and Millard, R. C., Jr., Measurements of vertical fine structure in the Sargasso Sea, *J. Geophys. Res.*, 80, 314, 1975.

124. Johnson, C. L., Cox, C. S., and Gallagher, B., The separation of wave-induced and intrusive oceanic finestructure, *J. Phys. Oceanogr.*, 8, 846, 1978.

125. Gordon, A. L., Georgi, D. T., and Taylor, H. W., An arctic Polar Front Zone in western Scotia Sea — summer 1975, *J. Phys. Oceanogr.*, 7, 309, 1977.

126. Georgi, D. T., Fine structure in the Antarctic Polar Front Zone: its characteristics and possible relationship to internal waves, *J. Geophys. Res.*, 83, 4579, 1978.

127. Stern, M., The salt-fountain and thermohaline convection, *Tellus*, 12, 172, 1960.

128. Turner, J. S. and Stommel, H., A new case of convection in the presence of combined vertical salinity and temperature gradients, *Proc. Natl. Acad. Sci.*, 52, 49, 1964.

129. Turner, J. S., Salt fingers across a density interface, *Deep Sea Res.*, 14, 599, 1967.

130. Williams, A. J., III, Images of ocean microstructure, *Deep Sea Res.*, 22, 811, 1975.

131. Magnell, B., Salt fingers observed in the Mediterranean outflow region (34°N, 11°W) using a towed sensor, *J. Phys. Oceanogr.*, 6, 511, 1976.

132. Veronis, G., A finite amplitude instability in thermohaline convection, *J. Mar. Res.*, 23, 1, 1965.

133. Martin, S. and Kaufmann, P., An experimental and theoretical study of the turbulent and laminar convection generated under a horizontal ice sheet floating on warm salty water, *J. Phys. Oceanogr.*, 7, 272, 1977.

134. Turner, J. S., *Buoyancy Effects in Fluids*, Cambridge University Press, New York, 1973.

135. Miles, J. W. and Howard, L. M., Note on heterogeneous shear flow, *J. Fluid Mech.*, 20, 331, 1964.

136. Woods, J. D., *Turbulence and Transition from Laminar to Turbulent Flow in Stratified Fluids*, Cahiers de Méchanique Mathématique, Université de Liège, Liège, 1969, 57.

137. Orlanski, I. and Bryan, K., Formations of the thermocline step structure by large-amplitude gravity waves, *J. Geophys. Res.*, 74, 6975, 1969.

138. Frankignoul, C. J., Stability of finite amplitude internal waves in a shear flow, *Geophys. Fluid Dyn.*, 4, 91, 1972.

139. Thorpe, S. A., On the stability of internal waves, in *A Voyage of Discovery*, Angel, M., Ed., Pergamon Press, New York, 1977, 199.

140. Thorpe, S. A., On the shape and breaking of finite amplitude internal gravity waves in a shear flow, *J. Fluid Mech.*, 85, 7, 1978.

141. Legeckis, R., A survey of worldwide sea surface temperature fronts detected by environmental satellites, *J. Geophys. Res.*, 83, 4501, 1978.

142. Roden, G. I. and Paskausky, D. F., Estimation of rates of frontogenesis and frontolysis in the North Pacific Ocean using satellite and surface meteorological data from January 1977, *J. Geophys. Res.*, 83, 4545, 1978.

143. Johannessen, O. M. and Foster, L. A., A note on the topographically controlled oceanic polar front in the Barents Sea, *J. Geophys. Res.*, 83, 4567, 1978.

144. Kinder, T. H. and Coachman, L. K., The front overlying the continental slope in the Eastern Bering Sea, *J. Geophys. Res.*, 83, 4551, 1978.

145. Beardsley, R. C. and Flagg, C. N., The water structure, mean currents, and shelf-water/slope-water front of the New England continental shelf, *Mem. Soc. R. Sci. Liege*, 10, 209, 1976.

146. Voorhis, A. D., Webb, D. C., and Millard, R. C., Current structure and mixing in the shelf slope water front south of New England, *J. Geophys. Res.*, 81, 3695, 1976.

147. Flagg, C. N. and Beardsley, R. C., On the stability of the shelf water/slope water front south of New England, *J. Geophys Res.*, 83, 4623, 1978.

148. Simpson, J. H. and Hunter, J. R. Fronts in the Irish Sea, *Nature (London)*, 250, 404, 1974.

149. Simpson, J. H., Allen, C. M., and Morris, N. C. G., Fronts on the continental shelf, *J. Geophys. Res.*, 83, 4607, 1978.

150. Pingree, R. D. and Griffiths, K. S., Tidal fronts on the shelf seas around the British Isles, *J. Geophys. Res.*, 83, 4615, 1978.
151. Monin, A. S. and Yaglom, A. M., *Statistical Fluid Mechanics,* MIT Press, Cambridge, Mass., 1971.
152. Launder, B. E., Reece, R. J., and Rodi, W., Progress in the development of a Reynolds-stress turbulence closure, *J. Fluid Mech.*, 68, 537, 1975.
153. Launder, B. E., On the effects of a gravitational field on the turbulent transport of heat and momentum, *J. Fluid Mech.*, 67, 569, 1975.
154. Webster, C. A. G., An experimental study of turbulence in a density-stratified shear flow, *J. Fluid Mech.*, 19, 221, 1964.
155. Gibson, M. M. and Launder, B. E., Ground effects on pressure fluctuations in the atmospheric boundary layer, *J. Fluid Mech.*, 86, 491, 1978.
156. Kim, H. T., Kline, J. S., and Reynolds, W., The production of turbulence near a smooth wall in a turbulent boundary layer, *J. Fluid Mech.*, 50, 133, 1971.
157. Brown, G. L. and Roshko, A., The effect of density difference on the turbulent mixing layer, in AGARD CP-93, Vol. 23, Advisory Group for Aerospace Research and Development, NATO, London, 1971, 1.
158. Hinze, J. O., *Turbulence,* McGraw Hill, New York, 1959.
159. Fjörstoft, R., On the changes in the spectral distribution of kinetic energy for two-dimensional nondivergent flow, *Tellus,* 5, 225, 1953.
160. Kraichnan, R., Inertial ranges in two-dimensional turbulence, *Phys. of Fluids,* 10, 1417, 1967.
161. Manabe, S., Smagorinsky, J., Holloway, J. L., Jr., and Stone, H. M., Simulated climatology by a general circulation model with a hydrological cycle, *Mont. Weather Rev.*, 98, 175, 1970.
162. Rhines, P. B., Geostrophic turbulence, *Annu. Rev. Fluid Mech.*, 11, p. 401, 1979.
163. Rhines, P. B., The dynamics of unsteady currents, in *The Sea*, Vol. 6, Goldberg, E. D., McCave, I. N., O'Brien, J. J., and Steel J. H., Eds., Interscience, New York, 1977.
164. Kolmogorov, A. N., The local structure of turbulence in an incompressible viscous fluid for very large Reynolds numbers, *C. R. (Dokl.) Acad. Sci. URSS*, 30, 301, 1941.
165. Obukhov, A. M., On the distribution of energy in the spectrum of turbulent flow, *C. R. (Dokl.) Acad. Sci. URSS*, 32, 19, 1941.
166. Onsager, L., The distribution of energy in turbulence, *Phys. Rev.*, 68, 286, 1945.
167. Ozmidov, R. V., On the turbulent exchange in a stably stratified ocean, *Izv. Atm. Ocean. Phys. Ser.*, 1, 493, 1965.
168. Lumley, J. L., The spectrum of nearly inertial turbulence in a stably stratified fluid, *J. Atmos. Sci.*, 21, 99, 1964.
169. Townsend, A. A., On the fine scale structure of turbulence, *Proc. R. Soc. London, Ser. A.*, 208, 534, 1951.
170. Woods, J. D., Space-time characteristics of turbulence in the seasonal thermocline, *Mem. Soc. R. Sci. Liege,* 6, 109, 1973.
171. Monin, A. S. and Obukhov, A. M., Dimensionless characteristics of the turbulence in layers of the atmosphere near the earth, *Dokl. Natl. Acad. USSR,* 93(2), 2, 23, 1953.
172. Winant, C. D. and Browand, F. K., Vortex pairing: the mechanism of turbulent mixing-layer growth at moderate Reynolds number, *J. Fluid Mech.*, 63, 237, 1974.
173. Browand, F. K. and Wang, Y. H., An experiment on the growth of small disturbances at the interface between two streams of different densities and velocities, in Int. Symp. Stratified Flows, Novosibivsk, U.S.S.R., 1972. (Novosibirsk Proc. 491—498).
174. Stigebrandt, A., Vertical diffusion driven by internal waves in a sillfjord, *J. Phys. Oceanogr.*, 6, 486, 1976.
175. Townsend, A. A., Turbulent flow in a stably stratified atmosphere, *J. Fluid Mech.*, 3, 361, 1958.
176. Taylor, G. I., Internal waves and turbulence in a fluid of variable density, *Rapp. P. V. Reun. Cons. Int. Explor. Mer,* 76, 35, 1931.
177. Miles, J. W., On the stability of heterogeneous shear flows, *J. Fluid Mech.*, 10, 496, 1961.
178. Lumley, J. L. and Panofsky, H. A., *The Structure of Atmospheric Turbulence, Interscience,* New York, 1964.
179. Thorpe, S. A., Turbulence and mixing in a Scottish loch, *Phil. Trans. R. Soc. London Ser. A* 286, 125, 1977.
180. Pedersen, Fl. Bo., A Monograph on Turbulent Entrainment and Friction in Two-Layer Stratified Flow, Series paper 25, Institute of Hydrodynamics and Hydraulic Engineering, Technical University, Copenhagen, 1980.
181. McBean, G. A., Stewart, R. W., and Miyake, M., The turbulent energy budget near the surface, *J. Geophys. Res.*, 76, 6540, 1971.
182. Csanady, G. T., Mean circulation in shallow seas, *J. Geophys. Res.*, 81, 5389, 1976.
183. Stommel, H. and Leetmaa, A., The circulation on the continental shelf, *Proc. Natl. Acad. Sci. U.S.A.*, 69, 3380, 1972.

184. **Bowden, K. F., Fairbairn, L. A., and Hughes, P.,** The distribution of shearing stresses in a tidal current, *Geophys. J.,* 2, 288, 1959.
185. **Csanady, G. T.,** Frictional currents in the mixed layer at the free surface, *J. Phys. Oceanogr.,* 2, 498, 1972.
186. **Tennekes, H.,** The logarithmic wind profile, *J. Atmos. Sci.,* 30, 234, 1973.
187. **Garrett, J. R.,** Review of drag coefficients over oceans and continents, *Mon. Weather Rev.,* 105, 915, 1977.
188. **Kullenberg, G.,** On vertical mixing and the energy transfer from the wind to the water, *Tellus,* 28, 159, 1976.
189. **Saunders, P. M.,** On the uncertainty of wind stress curl calculations, *J. Mar. Res.,* 34, 155, 1976.
190. **Walin, G.,** Some observation of temperature fluctuations in the coastal region of the Baltic, *Tellus,* 24, 187, 1972.
191. **Shaffer, G.,** Conservation calculations in natural coordinates — cross-isohaline advective and diffusive salt fluxes near the coast in the Baltic, *J. Phys. Oceanogr.,* 9, 847, 1979.
192. **Csanady, G. T.,** The coastal jet conceptual model in the dynamics of shallow seas, in *The Sea,* Vol. 6, Goldberg, E. D., McCave, I. N., O'Brien, J. J., and Steele, J. H., Eds., Interscience, New York, 1977, 117.
193. **Deacon, G. Sir,** Ed., A discussion on ocean currents and their dynamics, *Phil. Trans. R. Soc. London Ser. A ,* 270, 349, 1971.
194. **Stern, M. E.,** *Ocean Circulation Physics,* Academic Press, New York, 1975.
195. **Veronis, G. and Stommel, H.,** The action of variable wind stresses on a stratified ocean, *J. Mar. Res.,* 15, 43, 1956.
196. **Holland, W. R. and Lin, L. B.,** On the generation of mesoscale eddies and their contribution to the oceanic general circulation. I. A preliminary numerical experiment, *J. Phys. Oceanogr.,* 5, 642, 1975.
197. **Orlanski, I. and Cox, M. D.,** Baroclinic instability in ocean currents, *Geophys. Fluid Dyn.,* 4, 297, 1973.
198. **Pedlosky, J.,** Finite-amplitude baroclinic disturbances in downstream varying currents, *J. Phys. Oceanogr.,* 6, 335, 1976.
199. **Robinson, A. and Stommel, H.,** The oceanic thermocline and the associated thermohaline circulation, *Tellus,* 11, 295, 1959.
200. **Robinson, R. and Welander, P.,** Thermal circulation on a rotating sphere, with application to the oceanic thermocline, *J. Mar. Res.,* 21, 25, 1963.
201. **Welander, P.,** An advective model of the ocean thermocline, *Tellus,* 11, 309, 1959.
202. **Welander, P.,** The thermocline problem, *Phil. Trans. R. Soc. London Ser. A,* 270, 415, 1971.
203. **Roether, W., Münnich, K. O., and Östlund, H. G.,** Tritium profile at the North Pacific (1969) GEOSECS Intercalibration station, *J. Geophys. Res.,* 75, 7672, 1970.
204. **Peterson W. H. and Rooth, C. G.,** Formation and exchange of deep water in the Greenland and Norwegian Seas, *Deep Sea Res.,* 23, 273, 1976.
205. **Jenkins, W. J. and Clarke, W. G.,** The distribution of ^{3}He in the western Atlantic Ocean, *Deep Sea Res.,* 23, 481, 1976.
206. **Gregg, M. C., Cox, C. S., and Hacker, P. W.,** Vertical microstructure measurements in the central North Pacific, *J. Phys. Oceanogr.,* 3, 458, 1973.
207. **Oakey, N. S. and Elliott, J. A.,** Vertical temperature gradient structure across the Gulf Stream, *J. Geophys. Res.,* 82, 1369, 1977.
208. **Schmitt, R. W., Jr. and Evans, D. L.,** An estimate of the vertical mixing due to salt fingers based on observations in the North Atlantic Central Water, *J. Geophys. Res.,* 83, 2913, 1978.
209. **Huppert, H. E. and Manins, P. C.,** Limiting conditions for salt fingering at an interface, *Deep Sea Res.,* 20, 315, 1973.
210. **Linden, P. F.,** Salt fingers in a steady shear flow, *Geophys. Fluid Dyn.,* 6, 1, 1974.
211. **Schmitt, R. W., Jr.,** Recent laboratory experiments with heat-salt fingers, *Eos,* 58 (Abstr.), 419, 1977.
212. **Tait, R. J. and Howe, M. R.,** Some observations of thermocline stratification in the deep ocean, *Deep Sea Res.,* 15, 275, 1968.
213. **Lambert, R. J., Jr., and Sturges, W.,** A thermohaline staircase and vertical mixing in the thermocline, *Deep Sea Res.,* 24, 211, 1977.
214. **Matthäus, W.,** Mittlere vertikale Wärmeaustauschkoeffizienten in der Ostsee, *Acta Hydrophys. Berlin,* 22, 73, 1977.
215. **Woods, J. D.,** On the role of the seasonal thermocline in primary production, *Deep Sea Res.,* 28, in press, 1981.
216. **Garrett, C. and Horne, E.,** Frontal circulation due to cabbeling and double diffusion, *J. Geophys. Res.,* 83, 4651, 1978.
217. **Williams, R. B. and Gibbs, C. H.,** Direct measurements of turbulence in the Pacific Equatorial Undercurrent, *J. Phys. Oceanogr.,* 4, 104, 1974.

218. **Gibson, C. H., Vega, L. A., and Williams, R. B.,** Turbulent diffusion of heat and momentum in the ocean, *Adv. Geophys.,* 18A, 353, 1974.

219. **Osborn, T. R.,** Estimates of the local rate of vertical diffusion from dissipation measurements, *J. Phys. Oceanogr.,* 10, 83, 1980.

220. **Sarmiento, J. L., Feely, H. W., Koore, W. S., Bainbridge, A. E., and Broecher, W. S.,** The relationship between vertical eddy diffusion and buoyancy gradient in the deep sea, *Earth Planet. Sci. Lett.,* 32, 357, 1976.

221. **Welander, P.,** Theoretical forms for the vertical exchange coefficients in a stratified fluid with application to lakes and seas, *K. Vetensk. o Vitterhets-Samhället, Göteborg, Ser. Geophysica,* 1, 1, 1968.

222. **Armi, L.,** Some evidence for boundary mixing in the deep ocean, *J. Geophys. Res.,* 83, 1971, 1978.

223. **Armi, L.,** Effects of variations in eddy diffusivity on property distributions in the oceans, *J. Mar. Res.,* 37, 515, 1979.

224. **Richardson, L. F.,** Atmospheric diffusion shown on a distance-neighbour graph, *Proc. R. Soc. London Ser. A,* 110, 709, 1926.

225. **Batchelor, G. K.,** The application of the similarity theory of turbulence to atmospheric diffusion, *Q. J. R. Meteorol. Soc.,* 76, 133, 1971.

226. **Okubo, A.,** Oceanic diffusion diagrams, *Deep Sea Res.,* 18, 789, 1971.

227. **Talbot, J. W., and Talbot, G. H.,** Diffusion in shallow seas and in English coastal and estuarine waters, *Rapp. P. V. Reun. Cons. Int. Explor. Mer,* 167, 93, 1974.

228. **Kullenberg, G.,** Investigations on dispersion in stratified vertical shear flow, *Rapp. P. V. Reun. Cons. Int. Explor. Mer,* 167, 86, 1974.

229. **Stommel, H. and Arons, A. B.,** On the abyssal circulation of the world ocean. V. The influence of bottom slope on the broadening of inertial boundary currents, *Deep Sea Res.,* 19, 707, 1972.

230. **Veronis, G.,** Use of tracers in circulation studies, in *The Sea,* Vol. 6, Goldberg, E. D., McCave, I. N., O'Brien, J. J., and Steele, J. H., Eds., Interscience, New York, 1977, 169.

231. **Needler, G. T. and Heath, R. A.,** Diffusion coefficients calculated from the Mediterranean salinity anomaly in the North Atlantic Ocean, *J. Phys. Oceanogr.,* 5, 173, 1975.

232. **Richardson, P. L. and Mooney, K.,** The Mediterranean outflow — a simple advection-diffusion model, *J. Phys. Oceanogr.,* 5, 476, 1975.

233. **Lambert, R. B.,** Small-scale dissolved oxygen variations and the dynamics of Gulf Stream eddies, *Deep Sea Res.,* 21, 529, 1974.

234. **Schmitz, J. E. and Vastano, A. C.,** Entrainment and diffusion in a Gulf Stream cyclonic ring, *J. Phys. Oceanogr.,* 5, 93, 1975.

235. **Joseph, J. and Sendner, H.,** Über die horizontale Diffusion im Meere, *Dtsch. Hydrogr. Z.,* 11, 49, 1958.

236. **Schuert, E. A.,** Turbulent diffusion in the intermediate waters of the North Pacific Ocean, *J. Geophys. Res.,* 75, 673, 1970.

237. **Ewart, T. E. and Bendiner, W. P.,** Techniques for estuarine and open ocean dye dispersal measurement, *Rapp. P. V. Reun. Cons. Int. Explor. Mer,* 167, 201, 1974.

238. **Walin G.,** A theoretical framework for the description of estuaries, *Tellus,* 29, 128, 1977.

239. **Pingree, R. D., Holligan, R. M., and Mardell, G. T.,** The effects of vertical stability on phytoplankton distributions in the summer on the north-west European shelf, *Deep Sea Res.,* 25, 1911, 1978.

240. **Wunsch, C.,** Internal tides in the ocean, *Rev. Geophys. Space Phys.,* 13, 167, 1975.

241. **Denman, K. L. and Miyake, M.,** Upper layer modification at ocean station Papa: observations and simulations, *J. Phys. Oceanogr.,* 3, 155, 1973.

242. **Richman, R. and Garret, C.,** The transfer of energy and momentum by the wind to the surface mixed layer, *J. Phys. Oceanogr.,* 7, 876, 1977.

243. **Krauss, W.,** Internal waves and mixing in the thermocline, Paper no. 57, 11th Conf. Baltic Oceanographers, Rostock, DDR, 1979, 709.

244. **Kullenberg, G.,** Observations of vertical mixing conditions in the surface and bottom boundary layers of the Baltic, Paper no. 10, 11th Conf. Baltic Oceanographers, Rostock, DDR, 1979, 197.

245. **Price, J. E., Mooers, C. N. K., and Van Leer, J. C.,** Observations and simulation of storminduced mixed-layer deepening, *J. Phys. Oceanogr.,* 8, 582, 1978.

246. **Stewart, R. W. and Grant, H. L.,** Determination of the rate of dissipation of turbulent energy near the sea surface in the presence of waves, *J. Phys. Geophys. Res.,* 67, 3177, 1962.

247. **Langmuir, I.,** Surface motion of water induced by wind, *Science,* 87, 119, 1938.

248. **Pollard, R. T.,** Observations and theories of Langmuier circulation and their role in near surface mixing, in *A Voyage of Discovery,* Angel, M., Ed., Pergamon Press, New York, 1977.

249. **Taylor, G. I.,** Dispersion of soluble matter in solvent flowing slowly through a tube, *Proc. R. Soc. London Ser. A,* 219, 186, 1953.

250. **Taylor, G. I.,** The dispersion of matter in turbulent flow through a pipe, *Proc. R. Soc. London Ser. A,* 223, 446, 1954.

251. **Bowles, P., Burns, R, Hudswell, R., and Whipple, R.,** Sea disposal of low activity effluent, in *Proc. 2nd Conf. Peaceful uses of Atomic Energy,* Vol. 18, International Atomic Energy Agency, Geneva, 1958, 376.

252. **Bowden, K. F.,** Horizontal mixing in the sea due to a shearing current, *J. Fluid Mech.,* 21, 83, 1965.

253. **Pritchard, D. W., Okubo, A., and Carter, H. H.,** Observations and theory of eddy movement and diffusion of an introduced tracer material in the surface layer of the sea, in *Disposal of Radioactive Wastes into the Seas, Oceans and Surface Waters,* Unipub., New York, 1966, 397.

254. **Carter, H. H. and Okubo, A.,** A Study of the Physical Processes of Movement and Dispersion in the Cape Kennedy Area, Ref. 65-2, Chesapeake Bay Institute, The Johns Hopkins University, Baltimore, 1965.

255. **Okubo, A.,** The effect of shear in an oscillatory current on horizontal diffusion from an instantaneous source, *Int, J. Oceanogr. Limnol.,* 1, 194, 1967.

256. **Evans, G. T.,** A two layer shear diffusion model, *Deep Sea Res.,* 24, 931, 1977.

257. **Kraus, E. B., Ed.,** *Modeling and Prediction of the Upper Layers of the Ocean,* Pergamon Press, New York, 1977.

258. **Niiler, P. P.,** One-dimensional models of the seasonal thermocline, in *The Sea,* Vol. 6, Goldberg, E. D., McCave, I. N., O'Brien, J. J., and Steele, J. H., Eds., Interscience, New York, 1977, 97.

259. **Kitaigorodskii, S.,** Review of the theory of wind-mixed layer deepening, in *Liege 10th Colloq. Ocean Hydrodynamics,* Vol. 28, Nihoul, C. J., Ed., Elsevier, Amsterdam, 1979.

260. **Kullenberg, G.,** Entrainment velocity in natural stratified vertical shear flow, *Estuarine Coastal Mar. Sci.,* 5, 329, 1977.

261. **Kato, H, and Phillips, O. M.,** On the penetration of a turbulent layer into stratified fluid, *J. Fluid Mech,* 37, 643, 1969.

262. **Thorpe, S. A.,** Experiments on the instability of stratified shear flows: miscible flows, *J. Fluid Mech.,* 46, 299, 1976.

263. **Fofonoff, N. P.,** Measurements of internal waves from moored buoys, in Ref. 68-72, Geophysical Fluid Dynamics Program, Course Lectures and Seminars, Vol. I, Woods Hole Oceanographic Institution, Woods Hole, Mass., 1968.

264. **Kullenberg, B.,** Interne Wellen im Kattegatt, *Sven. Hydrogr. Biol. Komm. Skr. Ny Ser. Hydrogr.,* 12, 1935.

265. **Smith, R. L.,** Upwelling, in *Oceanogr. Mar. Biol.,* 6, 7, 1968.

Chapter 2

MODELS OF DISPERSION

Gerrit C. van Dam

TABLE OF CONTENTS

I. INTRODUCTION

Before we can talk about *models of dispersion* it has to be known what is meant by dispersion and what is to be a model. Dispersion, as meant in the title, has to be understood in the general sense of spreading, but of course within the limitation of the subject of this book. A model of dispersion can be anything that presents an image of the physical phenomenon of spreading in the sea, more or less accurate or even largely simplified. The model can be physical itself, such as a hydraulic scale model. In that case, properties of interest, like concentrations, have to be measured in or at the model. Physical scale models do not play a great part in marine dispersion problems. Therefore the emphasis will be on nonphysical models, concepts involving reasoning and calculation. If the calculations are complicated or laborious, a computer will be needed. In such a case, the computer program is sometimes referred to as the model. If the basic concept of the model is a set of differential equations, the method of solving may become rather dominating and appear in the name of the model or the model type, e.g. "grid model", "finite difference model", "finite element model", "superposition model".

The nonphysical models are often referred to as *mathematical models.* One may object that this term obscures the correspondence with physical reality and theory. *Computational models* may be a slightly better term, but it does not meet the objection. If there is just a set of formulas and definitions but no specific computational procedure or program, one might speak about a *theoretical model.*

Any model must be based upon observation, but when constructed, a certain amount of generalization and hypothesis is usually unavoidable or wanted. Therefore, after the construction phase, often to be followed by modifications, experimental verifications are desirable and usually necessary.[1] Improvement of parameter values remains always possible; this may refer to mean values as well as to variations and fluctuations. Although these remarks are formulated mainly in terms of mathematical modeling, to some extent they apply to physical models as well.

Models may be divided in classes from many different viewpoints. These include number of dimensions, stationarity, tools (computer, physical model), mathematical method (space grid, superposition principle), problems dealt with (instantaneous source, continuous source), purpose (predictive, descriptive, instructive) and so on.

It will not be tried to draw up a multidimensional scheme containing all these points of view, neither will subdivisions be made cutting up the subject according to all possible combinations. Fortunately, not all combinations make sense (e.g., mathematical method does not apply to physical models), and some are unimportant. A main division is made between physical and nonphysical modeling of which the latter category will be considered first, being the most comprehensive one, in this topic.

Some remarks from the viewpoint of purpose will follow now. Dealing with pollution, the purpose of *prediction*[1] seems to be the most important, at least ultimately. Prediction can be a first step towards prevention, or at least control of pollution. As said above, a model often grows by steps; the first versions may not be good enough for predictions; they are stages of construction as well as research tools. Some models give rise to the development of other models, perhaps of a quite different kind. Models, possibly different in character, may complement each other, e.g., one providing certain boundary conditions for the other. A well-known example is the "near-field" physical scale model combined with a "far-field" mathematical model.

Even if it is not possible to arrive at quantitative predictions of some reliability, a model can give a better insight into the problem or into certain phenomena, so that a guide is obtained for measures to be taken or situations to be avoided. Predictive models cannot only be useful for design purposes and long-term measures. If they are sufficiently fast, they can be an aid in the case of calamities such as shipwreck causing oil spills or poisonous clouds of dissolved matter.

Dispersion of oil and other floating substances will be dealt with in Chapter 9. The present chapter is confined to dissolved matter or very fine suspensions and properties (like heat) or organisms (like bacteria) behaving very similar or identical to dissolved materials. Except in Section XII and parts of Section IX and XIII it will be assumed that the dispersing materials or properties are passive. This means that concentrations are so low that the water movement is not notably affected and, secondly, that the contaminant follows the movement of the water. The latter is not the case, for example, with particles with falling rates or buoyancy such that the vertical distribution is affected (Section XIII).

Kullenberg[2] has stated that the physical transport of pollutants in the sea is governed by the motion in the sea. This is even true for nonpassive contaminants, but in that case there is, by definition, some interaction between contaminant and sea motion. The influence upon the motion of the sea is usually rather minor and of local importance. The disturbed region is mostly a small part of the entire polluted area. The local disturbance, especially its degree of stratification, can be of practical interest. Modeling of this part of the field is difficult because of the complexity and generally three-dimensional character of the problem and a "near-field" physical scale model may be required. Except for the sections mentioned, the part on theoretical or mathematical models will be restricted to passive contaminants or to those parts of the polluted field in which water movements are not affected by the contaminant, the latter being sufficiently diluted by near field processes. Care has to be taken if models meant for simulation of passive spreading of pollutants are tested by comparing their results with natural patterns of spreading in which the tracer is not passive to a considerable extent, such as discharge of large rivers into the sea or Mediterranean waters spreading in the Atlantic. Density effects may have noteworthy influence upon current patterns which in turn may lead to a different way of spreading than would be the case with passive materials. Since estuaries are separately dealt with in Chapter 8, little is said in this chapter about estuarine models, especially the one-dimensional and layered types.

II. THE VELOCITY FIELD

In the case of passive contaminants, the velocity field of the water mass is the unaffected natural field. If this field would be known in sufficient detail, no further physical knowledge would be required to predict dispersion patterns. Any passive particle follows, by definition, the local velocity at its successive positions. It is not necessary to include its Brownian or molecular motion, since the displacements thereby are negligible with respect to the ubiquitous and complex "macroscopic" water motion in the sea. For all practical purposes, it would be no problem if only averages of the velocities over a certain volume (say, 1ℓ or even a cubic meter) and certain time intervals (say, of 10 s) would be known. To compute the effect of a continuous release, it would "only" be needed to make a kinematic computation, releasing a particle, say every second, and computing the path of all particles in small steps of time and space, in accordance to the grid size of the given velocity field. After about 2 weeks (in the "prototype"), a million positions would have been computed. In spite of the discreteness of the particles and their finite number, for practical purposes the concentration field at that moment would be known in sufficient detail. Carrying on sufficiently long would reveal how long it takes to reach a steady mean in various parts of the field and how large fluctuations and periodic changes are around this mean.

One of the purposes of this imaginative exercise is to show that *diffusion* (molecular or other) has not entered the computation in a quantitative way and qualitatively only at the moments the "concentration" was reviewed. At those moments it was assumed that limited numbers of particles in relatively small volumes represent a more or less even distribution of a much greater number of molecules of the real pollutant. If more exact estimates are needed, it only means that smaller steps in time and space have to be taken or more particles per second have to be released, which is no objection for an imaginary operation.

The above represents a category of models that was not mentioned in the foregoing; we might call it an imaginative or a *thought model*. It cannot be used for actual computation but its purposes may be gain of insight and elucidation of certain problems, and it may be a basis for setting up more real models.

In our case the thought model illustrates that diffusion concepts and similar artifacts, so often met in mathematical modeling of transport processes, only serve to substitute our lack of knowledge of the details of the velocity field or to compensate the limitations of our kinematic computing capacity. The idea that diffusion can be left out if velocities are known in sufficient detail is sometimes illustrated to some extent in actual applications of grid models. Leendertse et al.[3-6] have computed coliform distributions in Jamaica Bay on a rectangular two-dimensional grid on the basis of (horizontal) water velocities computed on the same grid. The strong tidal movement in the capricious system of shoals and channels and the drying and flooding of banks and shoals give rise to a very complicated and rapidly changing velocity pattern. These characteristics of the velocity field show clearly on the grid of computation, although it is not extraordinarily fine (78×61 squares of 500×500 ft^2). The fact that a good reproduction of measured concentrations could be obtained without caring much about diffusion (coefficients could be varied by a factor of ten or more without influencing the results), illustrates to some extent the reasoning of the preceding paragraphs. "Sufficient detail" in the velocity field apparently means not necessarily an impossible number of grid points and an enormous number of time steps. One may observe that the indifference of the computation for diffusion coefficients indicates that a considerable amount of numerical diffusion (Section VII) was present in the computation. However, the fact that a good agreement with observations could be obtained demonstrates that numerical diffusion, in this case, did not cause serious errors. It is possible that it was just of the right order to complement the finite difference approximation of the velocity field.

Since the example seems very encouraging, it is necessary to point out that a similar result in an area of the same size in the open sea will be much more difficult to obtain. In Jamaica Bay, all pronounced details of the velocities in the entire area are strongly governed by very well-known boundary conditions: the complex topography and the water levels at the narrow entrance of the bay. Knowing the relevant boundary conditions for an area in the open sea with an equivalent accuracy is probably much more difficult. Further one should realize that in the concerned case a period as short as a few tidal cycles was sufficient for the simulation. For long-living substances in the open sea, periods of several weeks and longer have to be simulated. At the present state of computer technology, this is still a serious problem. Computational techniques have been proposed to meet this objection.[7] These solutions remain within the scope of finite difference approximations of transport phenomena. In a different approach,[8] also starting from a complete velocity field on a space grid, one uses the (interpolated and time-dependent) velocity field for computing the paths of a (large) number of particles. The details of the velocity field *not* present in the computed field but needed for obtaining the required detail in the concentration pattern are represented by extra velocity components of a stochastic character but with suitable statistical properties (Section VIII). In some cases, like instantaneous sources studied from the viewpoint of relative dispersion (Section IV), no other than proper statistical knowledge of the velocity field would even be enough. For continuous sources, at least some systematic characteristics of the velocity field have to be known.

The computation of velocity fields, although often combined with transport simulation within one model or program, will be considered to be beyond the scope of our subject, except in the section on nonpassive contaminants. For *passive* contaminants the computation of velocity fields and of transport of pollutants can be performed entirely separately; there is no interaction. But also in this case the two computations are not mutually independent; the velocity computation is independent of the transport calculation, but the transport depends upon the velocities and usually the velocity data are obtained from some kind of computation, often by means of a sophisticated mathematical flow model.

III. TURBULENCE AND TURBULENT DIFFUSION

It is difficult to avoid these terms in the context of pollutant transport. Turbulence is a type of flow of which statistical characteristics can be given but no actual velocities at specific times and positions can be predicted.[9-12] After Hinze,[9] " . . . turbulent fluid motion is an irregular condition of flow in which the various quantities show a random variation with time and space coordinates, so that statistically distinct average values can be discerned."

Sometimes it is possible to distinguish rather clearly between the systematic, predictable, or repeatable part of a velocity field U, V, W and the "turbulent" part u′, v′, w′, the complete picture being given by

$$u(x,y,z,t) = U(x,y,z,t) + u'(x,y,z,t)$$

$$v(x,y,z,t) = V(x,y,z,t) + v'(x,y,z,t) \tag{1}$$

$$w(x,y,z,t) = W(x,y,z,t) + w'(x,y,z,t) \quad \text{(notations as in Chapter 1)}$$

In a clear-cut case, it is indeed possible to make a *cut* between the two parts by proper averaging procedures. In such a case (u′, v′, w′) does not only consist of variations of higher frequency than (U,V,W) does, but there is even a gap between the two frequency ranges. In many cases, especially in the sea, the separation is not so clear. In the sea, irregular motions are observed with periods much longer than, for example, the main tidal periods. In theory, it does not matter that the frequency ranges of systematic and "turbulent" movements overlap. A separation can be made by taking ensemble averages over a sufficiently great number of sufficiently long periods. In practice one could do this for a small number of fixed points such as light vessel locations, but never for an entire sea (e.g., the North Sea). There are, however, also more fundamental problems with turbulence and especially turbulent diffusion in the sea, as we will see later.

Anyhow, turbulence and turbulent diffusion are widely used terms in the connection of pollutant dispersion (an entire book has been published under the title *Turbulent Diffusion in the Environment*[13]). So at least some attention has to be paid to these notions. Let us consider again the case that systematic and "random" velocities are well separated in terms of frequencies. The transport equation of a passive contaminant (concentration c)

$$\frac{\partial c}{\partial t} + u\frac{\partial c}{\partial x} + v\frac{\partial c}{\partial y} + w\frac{\partial c}{\partial z} = \kappa\left(\frac{\partial^2 c}{\partial x^2} + \frac{\partial^2 c}{\partial y^2} + \frac{\partial^2 c}{\partial z^2}\right) \tag{2}$$

(note the formal identity with the equation for salt Chapter 1; κ = coefficient of molecular diffusion) can be written in terms of Equation 1 if we write $c = \overline{c} + c'$ similar to $U + u'$, etc. By assumption it is possible to take averages over such a time T, that $u' = v' = w' = 0$ but without affecting the slower variations, contained in U, V, W. Assuming in compressibility ($\partial u/\partial x + \partial v/\partial y/ + \partial w/\partial z = 0$) and applying "Reynolds rules" for the various averages (for a more detailed discussion one may refer to Okubo[14]), we obtain

$$\frac{\partial \overline{c}}{\partial t} + U\frac{\partial \overline{c}}{\partial x} + V\frac{\partial \overline{c}}{\partial y} + W\frac{\partial \overline{c}}{\partial z} = -\frac{\partial \overline{u'c'}}{\partial x} - \frac{\partial \overline{v'c'}}{\partial y} - \frac{\partial \overline{w'c'}}{\partial z}$$

$$\tag{3}$$

$$+ \kappa\left(\frac{\partial^2 \overline{c}}{\partial x^2} + \frac{\partial^2 \overline{c}}{\partial y^2} + \frac{\partial^2 \overline{c}}{\partial z^2}\right)$$

Suppose one may write (isotropic and homogeneous case)

$$-\frac{\overline{u'c'}}{\frac{\partial \overline{c}}{\partial x}} = -\frac{\overline{v'c'}}{\frac{\partial \overline{c}}{\partial y}} = -\frac{\overline{w'c'}}{\frac{\partial \overline{c}}{\partial z}} = K \qquad \text{(constant)} \tag{4}$$

then Equation 3 becomes formally identical with Equation 2, with $K + \kappa$ instead of κ and averages of velocities and concentrations instead of the original values. If no isotropic and homogeneous K is assumed, Equation 4 must be replaced by

$$\overline{u'c'} = -K_{xx}\frac{\partial \overline{c}}{\partial x} - K_{xy}\frac{\partial \overline{c}}{\partial y} - K_{xz}\frac{\partial \overline{c}}{\partial z}$$

$$\overline{v'c'} = -K_{yx}\frac{\partial \overline{c}}{\partial x} - K_{yy}\frac{\partial \overline{c}}{\partial y} - K_{yz}\frac{\partial \overline{c}}{\partial z} \tag{5}$$

$$\overline{w'c'} = -K_{zx}\frac{\partial \overline{c}}{\partial x} - K_{zy}\frac{\partial \overline{c}}{\partial y} - K_{zz}\frac{\partial \overline{c}}{\partial z}$$

If the principal axes of the tensor $\vec{\overrightarrow{K}}$ coincide with x, y, and z, only the diagonal values K_{xx}, K_{yy} and K_{zz} are unequal to zero and the equation for "turbulent diffusion" (plus advection) may be written ($\kappa \ll K_{x,y,z}$):

$$\frac{\partial \overline{c}}{\partial t} + U\frac{\partial \overline{c}}{\partial x} + V\frac{\partial \overline{c}}{\partial y} + W\frac{\partial \overline{c}}{\partial z} = \frac{\partial}{\partial x}\left(K_x\frac{\partial \overline{c}}{\partial x}\right) + \frac{\partial}{\partial y}\left(K_y\frac{\partial \overline{c}}{\partial y}\right)$$

$$\tag{6}$$

$$+ \frac{\partial}{\partial z}\left(K_z\frac{\partial \overline{c}}{\partial z}\right)$$

(Remember that decay and source terms have been omitted; they do not affect the line or reasoning.)

It can be made plausible and has also been empirically demonstrated that, generally speaking, this turbulent diffusion concept holds well as long as the initial assumption about the frequency ranges is fulfilled. Unfortunately, in the sea, at least for the three-dimensional and (horizontally) two-dimensional case, the assumption is incorrect.

In the sea, in the horizontal direction, irregular motions are found up to very large scales (100 km and even more). This means that such an element of the motion, often called "eddy", possesses internal coherence over its entire extent, its scale, but appears and disappears at irregular times and variable positions. This irregularity, especially for the large (horizontal) eddies does not mean that they are essentially unpredictable in the sense that they would not appear in a numerical hydrodynamic model if correct detailed boundary conditions would be known. The irregularity of these motions may however be an indication that they are very sensitive for small changes in boundary conditions. Beside these changeable eddies, there can be at the same time more or less stable flow patterns, near complex rigid horizontal boundaries as well as under the influence of a pronounced bottom topography,[1,15-17] in a "frozen" picture looking precisely the same as the more changeable flow patterns. In a statistical treatment of dispersion, it is not possible to distinguish between these two types of motion. This means, that on the basis of observations of dispersion alone, it is not possible to draw conclusions about the energy distribution of large scale "real" turbulence and more constant large scale eddy-like motions.

The existence of some kind of eddy motion at practically all scales relevant in most problems contradicts the assumption about the restricted frequency range of the random motions, at least when considering horizontal dispersion. The problem is *not* that K (or $\vec{K}$) may depend on time and place. Dependence upon flow velocity or local energy dissipation would be no objection for application of the concept. The problem is, that a K as defined by Equations 4 or 5 in the sea (and in similar situations) *grows* with the extent of a dispersing mass. "Eddies" that can initially be considered to belong to the regular velocity field, just displacing a patch of dispersing matter or not affecting it at all, will later, when the patch has grown, become a part of it and behave like fluctuations. They can be incorporated in K by gradually increasing the averaging period T, but as a consequence K becomes dependent upon the size of the patch. Such a generalized turbulent diffusion concept has indeed been widely used to describe dispersion in the sea, especially "patch diffusion". As to the latter, it is clear that indeed the scale of the phenomenon and the corresponding momentaneous K have to be well defined. To apply this generalized turbulent diffusion concept on more complex distributions, it is always necessary that in some way the distribution can be decomposed into a number of basic distributions (like those considered in Section IV), for example, successively proceeding from a continuous source, as considered in Section VI.

The generalized concept, just briefly described, does not provide a complete solution of the problem; it is just a *model*. This particular model concentrates upon the internal development of dispersing systems ("relative diffusion"). The operation which gradually shifts more components of the velocity field towards the fluctuating part (in fact by definition) also affects the "average" velocity field, gradually "smoothing" it more and more. Also the part that shifts to the fluctuations u' etc. is "smoothed" by the averaging operation in the definition of K, so that, at best, a description of dispersion processes in the form of a statistical average will be obtained.

The concept of a K dependent on cloud size can be related to patch diffusion theories (as touched in Section IV) and it can also be used in the superposition model type (section VI), although this is often not explicitly done. Finite difference transport models (section VII), based upon a fixed-space grid require K-values that may depend upon grid size, local velocity, vorticity, energy, etc., but not on patch size. Does this mean that these models cannot be any good for dispersion in the sea? This is not necessarily so. But it is important to realize what is done in the above terms. The size of the grid implies a rather arbitrary cut between velocity fluctuations and "regular" velocities. The latter, in this case, are the explicit values computed in the grid points. K, roughly speaking, corresponds with subgrid details of concentration and velocities. Solving an equation like Equation 6 on the grid will only provide a realistic result if the velocities U, V, . . . (these are here the grid point values) contain all relevant details of the larger scales ($>$ grid size), including stochastic components. It is indeed possible that the usual velocity fields as calculated for water motion simulation in the sea[4,5,18,19] are too smooth (lacking detail as well as stochastic variations) to provide proper dispersion on medium and large scales. This has to be tested by comparison of numerical simulations with known distributions. No relevant literature seems to be available as yet.

For certain steady distributions the scale effect can be represented by a *place*-dependent K. Describing a stationary "slender" plume[13] in such a way, K increases going from the thin end of the plume in the direction of the wider parts downstream.

One-dimensional treatment of vertical dispersion in the sea, with coefficients K_z, independent of the distribution of the dispersing matter has given satisfactory results in a number of instances.[2] The coefficients may depend on the depth. This is not in contradiction with Equation 4 or 5. One may doubt whether depth dependence or a scale effect is dominant.[2] A discussion of coefficients of vertical turbulent diffusion and observed values was given earlier in this book,[2] including the influence of stratification.

The definitions in Equations 4 and 5 of turbulent diffusion coefficients are rather formal. As observed by Okubo[14] we must for this reason not be surprised if values of components of the K-tensor are sometimes negative. For similar reasons it should not surprise if the generalized concept (K dependent on patch size) fails on certain details such as the (average) spatial distribution of a patch that originates from an instantaneous point source (Section IV). In a stationary velocity field with uniform K (only depending upon patch diameter), the spatial distribution should be Gaussian to satisfy the equation. It is clear that one cannot at all be sure that this is realistic.

Good results with the eddy diffusivity concept may be expected when a bound is set on eddy size and the dispersing system is large compared to this size. This can be the case in an estuary where no direct physical exchange in longitudinal direction is to be expected over distances larger than the tidal excursion, so that use of Equation 6 in one-dimensional form should be successful for phenomena which have indeed a one-dimensional character (such as salt intrusion) and a spatial extension which is large compared to the tidal excursion length. Consequently, tidal averages of concentrations should be used. K may be dependent on place and possibly on river discharge[20,21] and tide[21] (spring or neap). Satisfactory results have indeed been reported for (vertically) well mixed estuaries[20-25] (also see Chapter 8).

IV. ANALYTICAL DESCRIPTIONS OF INSTANTANEOUS POINT SOURCE RELEASE IN TWO DIMENSIONS

The instantaneous point source is an attractive concept for theoretical and experimental studies because of its simplicity; a less complicated injection cannot be imagined. Besides, more complicated cases can be considered to be superpositions of this simple source.

The study of instantaneous point sources is connected with the concept of *relative diffusion* (e.g., Csanady,[13] Chapter 4, Chatwin and Sullivan[26]). This refers to all descriptions where concentrations are studied in a frame of reference attached to the center of gravity of the diffusing cloud. This approach corresponds with what in the preceding section was called the *internal development* of the cloud. The movement of the center of gravity itself is left out of consideration in this conception, which is sometimes referred to as the Lagrangean approach, generalizing the concept of the Lagrangean view of single particle movement or fluid motion where the moving particle or fluid parcel is followed instead of looking at velocities in fixed points (Eulerian velocities). This way of looking at a cloud leads to a somewhat peculiar separation between "diffusion" and "advection" which was already touched in the preceding section. The "advection" corresponds to a (Eulerian) velocity which is obtained by averaging over an ever increasing volume, determined by the cloud growth.

For theoretical treatment in two (horizontal) dimensions a vertical line source may seem to be ideal, but there are some objections. In the prototype a vertical line source is difficult to realize. Further there is the consideration that pollution is often introduced in a relatively thin layer and not over the entire vertical. Interpreting experiments one must realize that an injection in a relatively thin layer of thickness h will initially give a slower horizontal spread than a vertical source of extension $H \gg h$. In the latter case the vertical shear over H (see section V) will work immediately from the beginning; in the other case the shear effect is gradually mobilized as vertical spread over H takes place. In the conventional double-logarithmic plot of the top value of mass per surface area vs. time, the difference between the two cases will show as a "running ahead" in time of the vertical line source. The time shift will increase (it is zero at the start from the two-dimensional point of view) until the concentration level C_A is reached (Figure 1) at which the material of the thin layer source has just spread over H. At that point one might consider the patches identical but there may be quite a difference in spatial structure (see below). Inasmuch as the patches are really identical, the behavior of C_{2max} will further be the same, i.e., the curves in Figure 1 would be parallel from C_A in a linear plot (t_s is conserved). No experimental data about the shift t_s were found in literature, but one should expect that t_s can be of the same order or longer than the time needed for vertical spreading from h to H.

The vertical velocity shear as the main agent of the shift works during a limited period mainly in one direction. This will cause an elongated shape of the patch as it is indeed shown in Figure 4.11 in Okubo,[14] where it is explicitly mentioned that a vertical line source was made. The pronounced elongation effect connected with vertical line sources will often be just another reason (beside the technical problems) not to apply vertical line sources in instantaneous release experiments. The strong anisotropy in the beginning will influence the further experiment during a long period and may be even more serious a drawback for theoretical interpretations than the gradual mobilization of vertical shear that occurs with a source of small vertical extent.

The above considerations are believed to belong in a chapter on modeling because they represent an important example of interaction between model concept and experimental approach. In this example, (two-dimensional) model and experiment are, in the initial phase, not fully reconcilable and a compromise has to be made. This applies especially to the model as a tool to obtain physical insight. If the same model is used for simulation of pollutant disposal in a relatively thin layer, there is no conflict between model and experiment. In such a case, the empirical results can, directly or after some "styling", be carried into the model, with no or little theoretical consideration. The variability of the phenomena is studied empirically by repeated experiments.

Of course it is convenient to express the space-time history after instantaneous release of a mass M in analytical form, with a connection to theory wherever possible. For fruitful application such a description must have sufficient empirical support.

If the diffusivities are independent of space and time, Equation 6 permits analytical solutions if the velocity field, U, V, W is uniform (it need not to be steady). In that case we can even eliminate the terms with U etc. by transformation to a coordinate system attached to the patch, which is transported as a whole and unaffected by the mean field. If we take the center of the new coordinate system (ξ_1, ξ_2, . . .) in the center of the patch, the solution of Equation 6 in the period ($t = 0$, $t = \infty$) following after an instantaneous injection (no other sources, no decay) reads (p = number of dimensions):

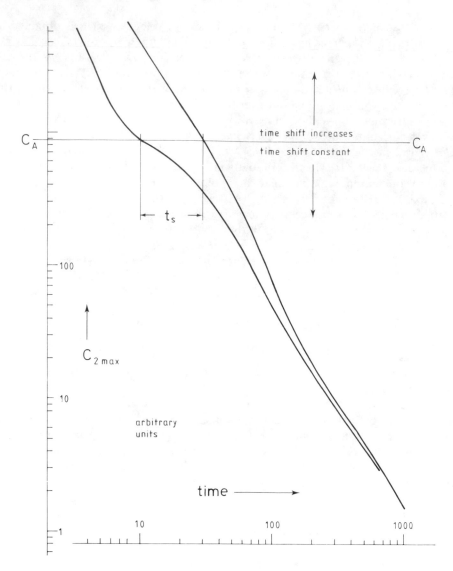

FIGURE 1.   Time shift in dilution process due to difference in initial conditions.

$$\bar{c}_p \,(\xi_i, t) = \frac{M}{(4\pi t)^{p/2} \,(\prod_i K_i)^{1/2}} \; e^{-\frac{1}{4t} \sum_i \frac{\xi_i^2}{K_i}} \qquad (t > 0) \qquad (7)$$

For example, in two dimensions:

$$\bar{c}_2 \,(\xi, \eta, t) = \frac{M}{4\,\pi t \sqrt{K_\xi K_\eta}} \; e^{-\frac{1}{4t} \left(\frac{\xi^2}{K_\xi} + \frac{\eta^2}{K_\eta} \right)} \qquad (8)$$

in which K_ξ, K_η need not to be different from K_x, K_y, provided that the axes of the new system are parallel to those of the original one. If valid, Equation 8 would in the sea be especially important for the case (ξ, η) = horizontal. The suffix of $\bar{c}$ reminds us that in the two-dimensional case, concentration (in terms of Equation 6) means mass per unit *surface* area, so for horizontal spreading in the sea: the vertically integrated concentration. If $\bar{c}_2$ is replaced by a volume concentration, a depth or a layer thickness must appear in the denominator of the right-hand side of the equation. Such a notation is rather common, quite often considering layer thickness or depth as a constant, while Equation 8 leaves it open whether this is so.

In the isotropic case, Equation 8 becomes

$$\bar{c}_2\,(r,t) \;=\; \frac{M}{4\pi Kt}\; e^{-\frac{r^2}{4Kt}} \tag{9}$$

in which the suffix of $\bar{c}$ is the only reminder of the number of space dimensions.

The underlying assumptions of Equations 7 to 9, uniform mean field and constant K or K_i, are generally unvalid in the sea as follows from the description of the velocity field in the preceding section. Therefore it should not be surprising that Equations 8 and 9 are not in agreement with most observations, especially as regards the *time* behavior. In most experiments at sea, $(c_2)_{max}$ decreases as t^{-2} to t^{-3} rather than t^{-1}. Temporarily a behavior $\propto t^{-1}$ may occur so that publications on short experiments sometimes report an exponent -1 for the overall description, but longer lasting measurements usually indicate larger (absolute) values of the exponent.[27-33]

Several authors[31,33,34] prefer a measure of the patch size instead of looking at the maximum concentration. For both ways of description, arguments can be given. If a "variance" σ^2 * or "standard deviation" σ* is chosen, it is necessary to measure and analyze the entire distribution. It can certainly not be taken for granted that the distribution is Gaussian, exponentional, or other. If this would be so, there would be a fixed relation between C_{max} and σ (at least if a σ is defined on the basis of an equivalent radius[35,28] r so that it does not depend upon axes). This fixed relation does not exist, even not within one experiment. In the extensive and accurate experiment RHENO 1965,[32] a marked change in type of distribution was observed, corresponding with a change in $(c_2)_{max}\,x\sigma^2$ by more than a factor of 2.

A next step to a better agreement with observations can be the application of the scale-dependent K in terms of the gradual shift of averaging time (and corresponding volume) as it was sketched in the preceding section. In these terms, proper momentaneous averaging means that velocity fluctuations on the scale of the patch are incorporated in K, because they disperse the patch, so that the velocities *within* the patch are smoothed out. This implies that the assumption of a uniform mean field becomes a fair approximation within the patch area, so that, strictly for the case of patch diffusion, the elimination of the terms with U etc. as applied in the derivation of Equation 7 now becomes justified under the provision that, correspondingly, K grows with patch size and consequently in time.

For simplicity, we consider the isotropic case so that polar coordinates can be used. In these coordinates, after elimination of the mean field, the equation (without sources or decay) becomes

$$K \left(\frac{\partial^2 \bar{c}}{\partial r^2} + \frac{p-1}{r}\,\frac{\partial \bar{c}}{\partial r} \right) = \frac{\partial \bar{c}}{\partial t} \tag{10}$$

---

* Except for the one-dimensional case, special definitions are necessary! See Equation 16 and later comments.

K is supposed to be place independent and therefore put before the differentiations with respect to r. An instantaneous point source of mass M at r = 0, t = 0 can formally be added to Equation 10 as a dirac-function $M\delta(r,t)$. The solution that we seek should satisfy Equation 10 for t > 0 and have $M\delta(r,t)$ as a limit for t → 0. We must further permit that K increases with patch size and thus with time. Let us allow for an extension of the "classical" case Equation 7 in such a sense that $(\bar{c}_p)_{max}$ decreases as $t^{-\beta}$ without the restriction $\beta = p/2$ as in Equation 7. In two-dimensional patch studies a fair approximation is often possible in such terms, with β = constant (and $\geqslant p/2$). Such a solution in the general p-dimensional form is obtained if we put

$$K \propto t^{\frac{2\beta}{p} - 1} \tag{11}$$

and it can be written

$$\bar{c}_p(r,t) = \frac{M}{(2\pi pKt/\beta)^{p/2}} e^{-\frac{\beta r^2}{2pKt}} \tag{12}$$

or, to make explicit the dependence on t:

$$\bar{c}_p(r,t) = Mc_0 t^{-\beta} e^{-\pi c_0^{\frac{2}{p}} t^{-\frac{2\beta}{p}} r^2} \tag{13}$$

For example:

$$p = 2 \rightarrow K \propto t^{\beta-1} \tag{14}$$

according to Equation 11.

The constant c_o can be expressed as a function of the proportionality factor of Equation 11. Further it can be proved that the above implies that

$$K \propto l^{2 - \frac{p}{\beta}} \tag{15}$$

if l is some measure of the patch "diameter", e.g., the root mean square value of the radius vector r of all "particles" of the patch:

$$\sqrt{\frac{\int_0^\infty r^2 \, \bar{c}(r)dV}{\int_0^\infty \bar{c}(r)dV}} \qquad \begin{array}{l} \text{(volume element} \\ dV = (2r)^{p-1}\pi dr \\ \text{for } p = 2 \text{ and } p = 3) \end{array} \tag{16}$$

which has been called σ_{rc} by Okubo.[36] It should be noted that various names (σ, σ_c, σ_r, σ_{rc}, σ_{rC}) and definitions of a σ for p > 1 (usually p = 2) are found in literature[33,36-39] and that the same name not always corresponds with the same definition.

In the generalized concept of Equations 12 and 13, K still plays its classical role in this sense that in each point on a fixed moment the local diffusive transport per second in direction x_i equals

$$-K_i \frac{\partial \bar{c}}{\partial x_i} \tag{17}$$

("gradient-type" diffusion). Therefore the spatial distribution remains Gaussian like it was found above for the case of time-dependent K. Solution of (6) with unequal values K_x, K_y, . . . gives essentially the same.

The normal distribution sometimes agrees well with experiments at sea, but it often does not.[28,32,40] This should not surprise since the local character of the gradient-type diffusion conflicts basically with the idea that the most effective "eddies" are not small compared to the size of the patch but of the same order of magnitude.

As to the time dependence of $\bar{c}_{max}$ or patch size, the agreement of observations with $t^{-\beta}$ (or $t^{\beta/2}$, respectively) is often rather good, also for relatively long periods, at least as a general trend.[14,31,33,41] The exponent β for the two-dimensional description of patch diffusion in seas and oceans usually lies between 2 and 3; the general trend over a long period and combined data of many different places indicates a β only little more than 2 (Figures 2 and 3).[14,34,41,42] The case $\beta = 3$ (for two dimensions) has been given much consideration for theoretical reasons, and on shorter ranges of time, empirical evidence for this value of β has been reported.[36,43,44] The result $\beta = 3$, or more general: $\beta = 3/2P$ should be expected in the case that the dispersion would be governed by the turbulence of a so-called inertial (sub)range[10,45-48] of stationary homogeneous turbulence. This conclusion can be obtained by dimensional analysis from the fact that the mutual dispersion of two particles in such an inertial range (for distances large compared to the dissipation scale) can only depend upon the energy dissipation per unit mass, ε, being the only parameter characterizing the turbulent motion.[12,46] In terms of a patch-size-dependent K, in an inertial range the velocity v_L characterizing the turbulent motion at scale L must vary as $L^{1/3}$ to obtain the necessary equal turbulent energy transfer at all scales. The scale-dependent diffusion coefficient K_L (dimension: velocity × length) must be proportional to $v_L \times L$, so it is proportional to $L^{4/3}$ in the inertial range. In patch diffusion, the scale L must be proportional to patch "diameter", characterized by a measure l (such as σ_{rc}). This is consistent with Equation 15 if $\beta = 3/2$ p.

In fact the turbulence spectrum in the sea is more complicated than a single inertial range.[49] Energy is injected at various scales and not just at the upper end (largest scale) of the range of interest. Further it may be doubted whether the properties of an inertial range are present[49] in the vast domain of large and very large *horizontal* "eddies" constrained by the relatively very small vertical extent of the water mass or mixed (upper) layer. Nevertheless there has been quite a bit of speculation[34,50-52] about inertial ranges and "cascades" of consecutive inertial ranges in the sea, up to the scales of hundreds or even a thousand kilometers. In any case, the motion in the sea has the general property that turbulence or at least eddy-like motions (having the same dispersive effect) are present from very small to very large scales, the latter usually exceeding the scale of phenomena of interest by far. Therefore it is understandable that in terms of scale-dependent diffusivity an approximation by

$$K \propto L^\gamma \qquad (\gamma \gtreqless 0) \qquad (18)$$

as a kind of generalization of the "$L^{4/3}$ law" often gives rather good results. As seen above, Equation 18 (essentially the same as Equation 15; l $\propto$ L) implies $(\bar{c}_p)_{max} \propto t^{-\beta}$ with $\gamma = 2 - \dfrac{p}{\beta}$, i.e., $\beta = p/2 - \gamma$. For the two-dimensional case

$$\gamma = 2 - \frac{2}{\beta} \quad \text{or} \quad \beta = \frac{2}{2 - \gamma} \qquad (p = 2) \qquad (19)$$

In these terms it makes no difference what linear measure l or patch size is used (like σ_{rc} or a multiple of it).

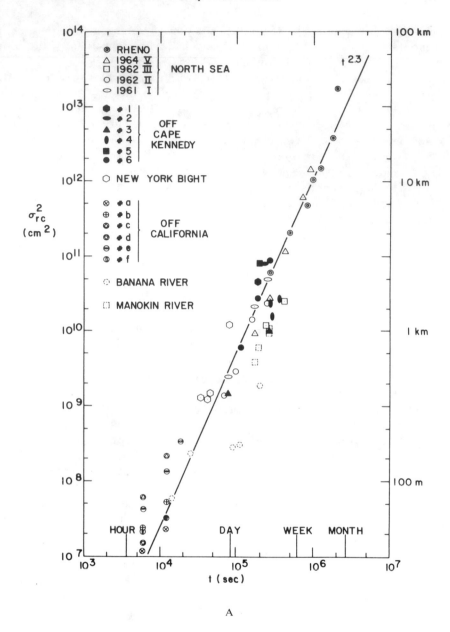

FIGURE 2.   (A) Patch growth for combined data from various places and conditions after Okubo[36] (1968). (B) "Apparent diffusivity" (vertical axis, cm² s⁻¹) vs. scale of diffusion after Okubo[44] (1971).

Also with arbitrary γ (and corresponding β), the above has the limitation of a normal distribution connected with the (local) gradient-diffusion conception. To meet this objection, several solutions have been proposed allowing for a K that also depends on the radius r (in isotropic terms). Surveys of this group of solutions can be found in Neumann and Pierson,[53] Okubo,[37] and Nihoul.[54] A subclass of these solutions is the one where the generalized coefficient of diffusion is $a \cdot r^m \cdot f(t)$. The possible physical meaning of such a coefficient is unclear. This objection also holds for the subclass discussed before (m = 0; f(t) a constant power of (t). Referring to Table 13.1 in Neumann and Pierson[53] for the other cases, we only pay some attention to the cases where,

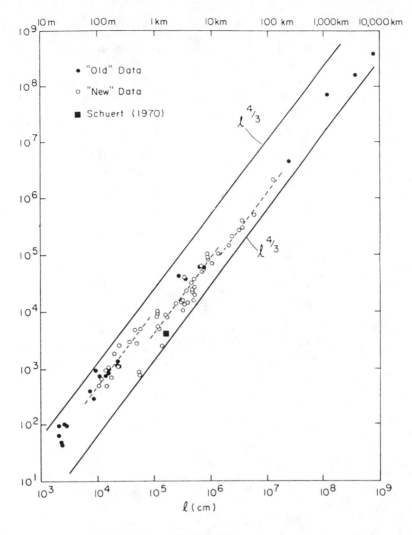

FIGURE 2B

in contrary to the above where K depended upon t only, K solely depends upon r. In literature we only find $K \propto r^m$ which can be considered an alternative way of expressing the scale dependency of the type in Equation 18. It implies that the differential equation cannot be written in the form of Equation 10: K must now remain within the (second) differentiation with respect to r (compare Equation 6). For the two-dimensional case the equation thus becomes:

$$\frac{\partial \bar{c}}{\partial t} = \frac{1}{r} \frac{\partial}{\partial r} \left(a\, r^{m+1} \frac{\partial \bar{c}}{\partial r} \right) \qquad \begin{array}{l} (p = 2) \\ ar^m = K \end{array} \qquad (20)$$

Joseph and Sendner[35] have treated the case m = 1 (giving the experimentally often rather well supported result $(\bar{c}_2)_{max} \propto t^{-2}$) and Ozmidov[55,56] the case m = 4/3 (corresponding with the "4/3 law" of the inertial range). Although the relation $K = ar^m$ is taken at each place and moment, in practice it functions indeed as a scale-dependent diffusion $\propto l^m$ because the dominant transports find place in the region of the largest gradients (approximately around $r = \sigma_{rc}$) which expands at the rate of patch size l ($\propto$

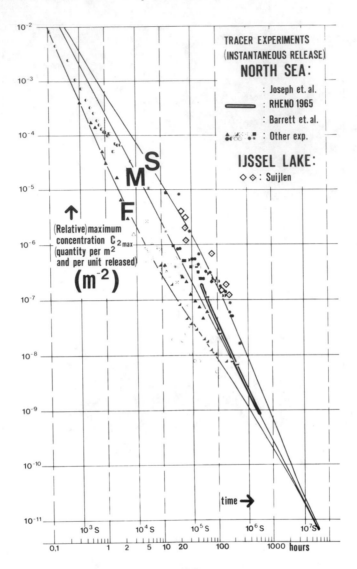

FIGURE 3.   Decrease of maximum concentration of patches for experiments in the North Sea at different locations and under various conditions.[41,42] IJssel Lake data[59] added for comparison. S, M, and F indicate "slow", "medium", and "fast" dilution.

σ_{rc}). The fact that this K at the same time tends to zero for $r \to 0$ and to infinity for $r \to \infty$ at all instants makes the approach physically just as doubtful as the case treated in the preceding where K is uniform at each instant but increases with patch size.

The respective results are (in two dimensions as published by the original authors):

$$\bar{c}_2(r, t) = \frac{M}{2\pi P^2 t^2} e^{-\frac{r}{Pt}}$$

("diffusion velocity" P = a in Equation 20)   (21)

(m = 1), Joseph and Sendner[35]

and

$$\bar{c}_2\,(r, t) = \frac{M}{6\pi \left(\dfrac{4}{9}a\right)^3 t^3}\; e^{-\dfrac{r^{2/3}}{\dfrac{4}{9}\,at}} \quad \begin{array}{l}\text{(a as in Equation 20)}\\[4pt] \left(m = \dfrac{4}{3}\right),\ \text{Ozmidov}^{55}\end{array} \qquad (22)$$

The t-dependence of $(\bar{c}_2)_{max}$ is for these cases $K \propto r$ and $K \propto r^{4/3}$ independent of t, identical with that obtained with $K \propto 1$ and $K \propto 1^{4/3}$, independent of r. The spatial distribution however differs from the Gaussian one. An example of the use of a Gaussian distribution not combined with a constant K is the formula proposed by Okubo and Pritchard[37,57,58] (Equation 12 or 13 with p = 2, β = 2).

It was remarked earlier that in terms of $K = K_L = L^\gamma$, $\gamma = 4/3$ (as for an inertial range) corresponds with a behavior of the characteristic velocity v_L as $L^{1/3}$, i.e., with a small but steady increase with scale and thus, for patch diffusion, with time. The case $\gamma = 1$ corresponds with a v_L independent of L and a t^{-2} behavior of c_{max} in two dimensions, which means that the patch size grows with a constant velocity as well. This is the parameter P in the solution of Joseph and Sendner. The same is true in one and three dimensions.

In experiments the agreement with any of the proposed spatial distributions varies from case to case and seems not to be related to the time dependence. Too much reliance on theoretical and in this case rather formal and physically untransparent solutions can run into trouble when interpreting data. Nihoul[54] analyzes the development in time of cloud shape(s) (in terms of radial symmetry) of three experiments in the North Sea and observes that (at least in two of the three cases) a gradual shift from a more or less exponential distribution to a Gaussian-like distribution seems to take place (which is opposite to many other observations like RHENO 1965). From the fact that Joseph and Sendner's theory combines an exponential distribution with $(\bar{c}_2)_{max} \propto t^{-2}$, and classical diffusion theory (constant diffusivity, Fick's law) a normal distribution with $(\bar{c}_2)_{max} \propto t^{-1}$, Nihoul concludes that in the experiments the diffusion rate, as reflected in the exponent of t, decreases. This conclusion entirely depends upon an absolute validity of both theories, also in such a sense that one assumes that a normal distribution, for example, cannot occur when $(\bar{c}_2)_{max}$ changes in some other way than $\propto t^{-1}$. The vast empirical material provides no basis for such a belief. A check of the data referred to by Nihoul learns that also in this particular case the time behavior does not agree with Nihoul's conclusion. In this example, the use of models has not been an aid or a guide for the interpretation of the data, but obscures the truth. Of course, this is not the fault of the models, but it is certainly not the only instance such a thing has happened and it may serve as a warning.

Repetition of instantaneous tracer experiments at one location can give considerable differences in results as to time behavior. On the other hand, the divergence between results from totally different locations is hardly greater, which in fact means that the process is rather consistent. This may be remarkable for a restricted region like the North Sea[41,42] (Figure 3), it is even more surprising in the well-known diagrams of Okubo[34,36,44] in which data from the entire world, from the ocean (surface layer), as well as from shallow seas and even great lakes, are combined (Figure 2). It strikes one that the scatter of the worldwide data is hardly larger than that in the North Sea data.

In spite of the fact that differences locally can be almost as large as the entire bandwidth, some systematic trends can be found. This mainly regards location; the role of weather conditions is more obscure and indirect. That there is an influence follows from the great variability at one location, but various factors seem to combine in a complex fashion, most likely with delayed effects.

As regards location, water depth, and intensity of tidal currents certainly play a role. In Figure 3, one group of data outside the North Sea has been included for comparison[41,59] (IJsselmeer 1971). In the IJsselmeer, there is no tide. The results are found at the upper edge of the data band.* Closest of these points are those from the regions relatively far from the Dutch coast where tidal amplitudes are comparatively small. The results of experiments near the coast (shallower waters, stronger tides, and more shear) are mainly found in the lower and middle part of the data band. This all refers mainly to the "medium" time scales. The diagram suggests strongly that the data converge at the left (upper) end (periods less than a few hours) as well as the right (lower) end (100 hr and more; length scales > 10 km). The latter can be understood by consideration of the fact that the influence of the tide, for example, becomes smaller and smaller as the patch grows, the tidal excursion being limited. It finally means no more than a minor vibration within the extended concentration field. A similar argument holds for weather variations, but mainly in time instead of space. Storms of limited duration and other temporary deviations from average conditions have less influence on patch size and concentration as the patch grows larger. This is quite similar to the decreasing "sensivity" of the patch to the other agents, constant or intermittent, taking care of its gradual growth: when time proceeds, ever-increasing periods of time are needed to obtain the same dilution factor. If $\bar{c}_{max} \propto t_\beta$ (β constant), which is always a fair approximation of the general trend, equal dilution factors correspond to equal factors in time. For example, if in the period from 10 to 100 hr (after point release) a dilution $1/n$ is obtained, the time required for the next dilution of $1/n$ is the period from 100 to 1000 hr, ten times longer than the preceding one. This holds for constant K as well as for K increasing according to Equation 18 with $\gamma > 0$. Neglect of this fact sometimes leads to overestimation of the diluting power of the sea.

In the seas, energy is supplied from outside (wind, tides, and oceanic circulations due to density differences) at various scales. Some of this energy is directly passed on to small scales by bottom friction, especially in shallow seas. The relatively small-scale turbulence distracted by bottom friction or similar mechanisms from water movements on a much larger scale, diminishes the amount of energy that is transmitted to small scales via intermediate scales. The energy of the small-scale turbulence itself may be passed on by a (three-dimensional) inertial subrange. It is unsure whether the large-scale energy that is not lost by boundary friction is passed on in the way of a (two-dimensional!) inertial range. In any case, several of these energy cascades as far as they exist are present at the same time. A description of the relevant scales and the possible inertial subranges has been given by Ozmidov[51,56] (Figure 4). Other authors have sketched the possible relation with observed behavior of $(c_2)_{max}$ or $\sigma_{(rc)}$. The t^{-3} behavior of $(c_2)_{max}$ would hold in limited-scale ranges, and consequently the overall behavior should correspond with a smaller exponent β (Figure 5).[34] Okubo's illustration shows that the data certainly do not contradict the idea. A similar interpretation of North Sea data has been reported earlier, but the present, largely extended set of data (Figure 3), although not contradictory with the general idea, does not give any indication of discrete levels of energy supply. It could be that the theory holds better for the mixed (upper) layer of the deep ocean where "bottom" friction is virtually absent than in a shallow sea where large eddies most likely lose more energy by bottom friction than by smaller eddies feeding upon them.

The essential feature responsible for the general time behavior sketched above is the presence of eddies or eddy-like motions at all scales, at least in the concerned experiments. The "largest eddies" are apparently not reached in artificial dye tracer experi-

---

* It is remarkable, though, that Murthy[60] comparing Lake Ontario and oceanic data reports a difference in the opposite sense.

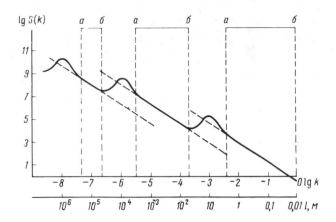

FIGURE 4. Tentative diagram of energy density distribution over oceanic motions of different scales, according to Ozmidov.[51,56] a—b: zones where 5/3 power law applies (corresponding to $L^{4/3}$ law for diffusion).

ments. Experiments with distinct markers commonly do not reach larger scales either. However, most recently an experiment has been reported[61] in which marker buoys in the Pacific Ocean were followed by a satellite, up to mutual distances of several hundreds of kilometers. It was found that when the cluster had reached a scale of about 250 km, the process slowed down and thereafter went on in the way of classical (Fickian) diffusion (constant K) in a homogeneous velocity field. Apparently, the scale of the largest eddies in this area was surpassed in this experiment and compared with the size of the buoy cluster, the velocity field became homogeneous. Its fluctuations or eddies became small compared with "patch" size and only contributed to the, now constant, eddy diffusivity K.

As remarked earlier, not only time behavior but also spatial distributions are rather variable[32,37] according to the empirical data. It should be observed that there are less reliable data on spatial distribution than on time behavior of maximum concentration or patch size. Often the data are not sufficiently accurate to decide between theoretical distributions[40] or the data do not fit any of the proposed formulas.

Still, a better theoretical basis for the average spatial distribution remains most desirable. Especially certain cases of superimposed distributions (next section) are rather sensitive to the basic distribution function applied (Figure 21C). On one hand, this implies the possibility to come to a first choice on the basis of empirical data from continuous sources, but this will never indicate one particular function with certainty.

Solutions based upon gradient diffusion concepts as discussed above will never lead to a fundamentally sound concept, simply because they all neglect the essential feature of exchange over an entire range of distances due to the existence of a broad spectrum of eddies. Consequently they also neglect the time lags connected with a more general exchange picture.[48,62] A concept accounting for these basic facts could still be rather general by allowing various approximations of the spectral distribution of "turbulent" energy which can be determined empirically. A very interesting approach in this direction has been made by Schönfeld[62] at the IUGG symposium of September 1961. His concept of "integral diffusivity" accounts for the fact of exchange over all distances, by a spectral representation of the "diffusion coefficient", thus integrating over the various distances while accounting for the corresponding differences in concentration. Applying this idea to the case of an instantaneous point source, Schönfeld succeeds in deriving analytical expressions for the distribution in space and time for some of the

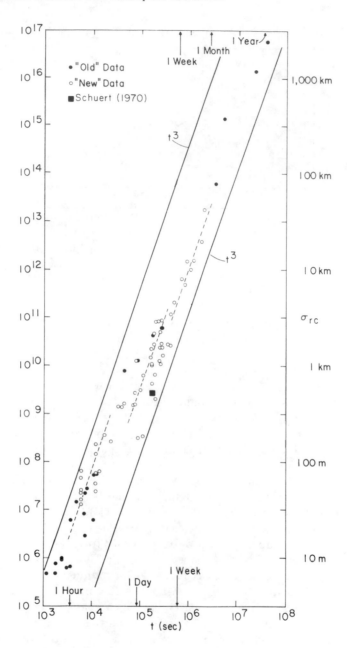

FIGURE 5.   Local fit to t^3 law according to Okubo.[44] Vertical axis σ^2_{rc} in cm².

most common analytical forms of the spectrum of turbulence, including the case of an inertial subrange. The solutions have the expected time behavior of c_{max}, i.e., the same that is predicted by the heuristic methods with generalized gradient diffusion (discussed earlier), but they further exhibit a spatial character that is indeed more pointed than the Gaussian, in agreement with some of the empirical evidence. However, Schönfeld remarks that his first solutions should be corrected for the fact that initially he has not accounted for retardation effects which certainly will occur in the exchange over increasing distances. Although Schönfeld gives some suggestions how to account for this complication and finds a number of "corrected" solutions, he does not succeed in solving the problem unambiguously, leaving us with a number of possible distributions including the Gaussian, dependent upon the "degree" of retardation. Unfortunately, no further progress has been reported although the approach has a better theoretical basis than most of the other concepts.

Below, some of Schönfeld's results are presented for the case of two dimensions and a constant "diffusion velocity" W (compare P in Joseph and Sendner's formula (Equation 21)) (Schönfeld's notations; s = concentration).

$$s = \frac{C}{\pi^2 (Wt)^2} \, e^{-\frac{r^2}{\pi (Wt)}} \qquad \text{normal distribution} \qquad (23)$$
$$\text{(strong retardation)}$$

$$s = \frac{C}{\frac{8}{3} \pi (Wt)^2} \left[1 + \frac{r^2}{4 (Wt)^2} \right]^{-\frac{5}{2}} \text{(retarded)} \qquad (24)$$

$$s = \frac{C}{2\pi (Wt)^2} \left[1 + \frac{r^2}{(Wt)^2} \right]^{-\frac{3}{2}} \quad \text{(unretarded)} \qquad (25)$$

In computations of dispersion in the North Sea with the two-dimensional superposition model of Rijkswaterstaat[29,63] (Section VI), the intermediate spatial distribution of Equation 24 is usually applied because this distribution agrees with a larger percentage of the dye experiments performed so far than the normal distribution does.[40]

As a matter of fact, Monin[64] (see also Monin and Yaglom[46]) was the first to propose a spectral representation for the diffusion coefficient, and consequently an integral equation, for turbulent diffusion.

A recent extension on Monin's results was presented by Pasmanter.[48] Like Monin, he bases his work on Kolmogorov's ideas of an inertial range and then uses dimensional arguments in order to derive the general expression for turbulent diffusion, including retardation effects which were left aside by Monin. The final expression obtained by Pasmanter contains a number of parameters which have to be determined from experiments. In some particular cases, e.g., the stationary distribution due to a permanent point source, there are no free parameters in the final solution; these cases could be used for experimental checking of the correctness of the theory. As it was said in the foregoing, the validity in two-dimensional systems of Kolmogorov's idea of an inertial subrange is uncertain, both theoretically and experimentally.

Another recent presentation of a spectral diffusivity concept was given by Berkowicz and Prahm.[65,66]

This section mentioned comparing theoretical and measured distributions without indicating how this should be done. The dash over c in the formulas left over from the derivation of the equations reminds us that we should take averages over ensembles of experiments. In practice this is not done; the number of experiments is always small. To some extent it is done for c_{max}; namely, when results of several experiments (usually of quite different locations as well) are plotted in one graph (Figures 2, 3 and 5) and the overall behavior is regarded. As to the spatial distribution, a certain averaging is possible within one observed patch. This is the common technique, maybe first introduced by Joseph and Sendner[28,35] in the form of equivalent radii: surfaces within isoconcentration lines are considered as if they were circles with the same surface area, so that the distribution can be compared with the theoretical formulas in their isotropic form (Equations 9, 12, 13 and 21 through 25).

As it was said earlier, due to the limited sampling path the accuracy of *in situ* mea-

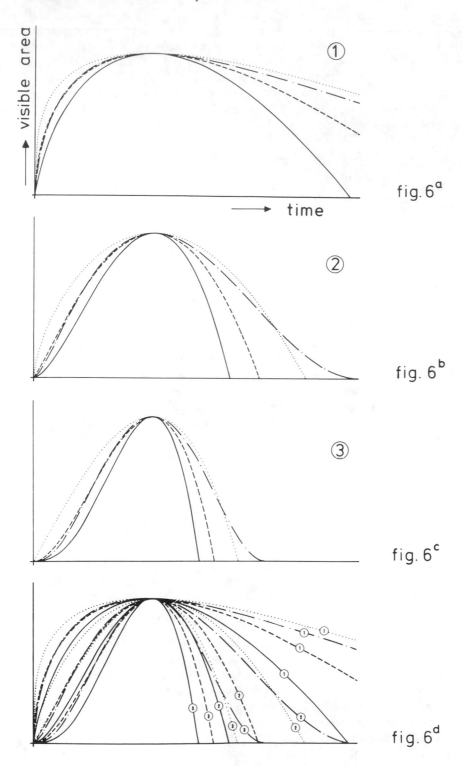

FIGURE 6.   Visible area (or area within some other threshold concentration) as a function of time for theoretical distributions — Gauss distribution; — Exponential distribution (Joseph and Sendner[35]); --- Schönfeld[62], formula (24) in this chapter; . . . Schönfeld[62], formula (25) in this chapter Curves have been scaled in such a way that the position of the point representing the maximum area is the same for all curves. (a) $c_{2\ max} \propto t^{-1}$, (b) $c_{2\ max} \propto t^{-2}$, (c) $c_{2\ max} \propto t^{-3}$, (d) all curves combined.

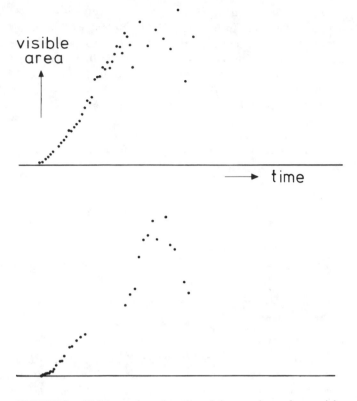

FIGURE 7. Visible area as a function of time as observed on aerial photographs of dye patch in the IJssel Lake[59] (upper) and the North Sea (lower graph).

surements (by moving ships[28,32,40,67]) is often insufficient or hardly sufficient[32,40] for deciding which theoretical spatial distribution fits best with the data. For small patches (some tens of meters) it is even quite impossible to obtain distribution data by ship and for medium size patches (some hundreds of meters), ships disturb the patch too much for a reliable measurement. For these small and medium sizes, aerial photography of (dye) patches is a means to obtain relevant data. It is difficult to obtain reliable distributions from densimetric analysis[68] of the photographs, but at least some insight can be obtained by means of analyzing the time behavior of the visibility boundary.[69] This behavior (indicated as "threshold curves" by Okubo[37,70]) is rather sensitive to the spatial distribution (Figure 6). The measurement of the relating areas can be rather accurate although the accuracy decreases with increasing area as a consequence of the decreasing contrast (Figure 7). Another complication is the fact that the curves also depend upon the time behavior (Figure 6) of c_{max} (say the value of β, which may be not quite constant as well!).

Recently, Suijlen[71] has pointed out that the limited information obtained by using a certain visibility threshold (e.g., as connected with a particular color film), neglecting the structure within the boundary, can be extended by a simple artificial variation of the visibility. Results so far favor the Gaussian distribution. The distribution function once established, the threshold technique can further be used for accurate determination of the time behavior of c_{max}. For these small scales, this is not possible with *in situ* techniques.

Apart from various time histories (maximum concentration, variance, standard deviation, thresholds, etc.), some type of diffusion coefficient remains an attractive parameter for characterizing (patch) diffusion processes, since it provides a measure of the kinematic process on specific moments, independent of preceding history, time passed after release, etc., provided that a *proper definition* is used. The possibility of maintaining the classical relation (Equation 17) for the diffusion coefficient (also in the generalized concept in the sense of scale-dependent diffusion), is cancelled as soon as the shape of the patch is non-Gaussian. It is desirable to have a definition of a generalized K that also applies for other distributions, e.g., for mutual comparison of observations. Of course, it whould be preferable to define this K in such a way that K coincides with the gradient concept according to Equation 17 when this applies. This can be done by taking a property of the "classical" K as a definition for the more general coefficient K_g. Such a definition[38] is

$$K_g \overset{def}{=\joinrel=} \frac{1}{2p} \frac{d\sigma^2(rc)}{dt} \tag{26}$$

which is consistent with Equations 10 or 17 and thus with Equations 12 and 13 if for σ or σ_{rc} the definition in Equation 16 is used. Some of the definitions in literature, such as

$$K = \frac{1}{4} \frac{d\sigma^2}{dt} \quad \text{for} \quad p = 2 \tag{27}$$

correspond to Equation 26, but unfortunately several other definitions do not, so that they are inconsistent with the gradient concept where it applies (Gaussian distributions). Sometimes the difference is just a constant factor (such as ½ instead of ¼ in Equation 27, although p = 2 and not 1). Sometimes, however, the difference also affects the way in which K depends upon time and patch size. This happens when Equations 26 or 27 are replaced by a formula (such as constant × σ^2/t) which only conforms with Equation 26 (possibly except for a constant factor) if σ^2 is a power function of time with constant exponent, and right from the beginning. This does not have to be the case at all.

The application of Equation 26 on anisotropic or irregular (measured) distributions can be realized by transforming the isoconcentration curves into equivalent circles as indicated above.

A definition like Equation 26 which depends upon a σ as defined by Equation 16 cannot always be applied, since for certain distributions (e.g. Equation 25), the numerator in Equation 16 is infinite. One may doubt whether such a distribution will ever be measured and, if so, whether it still makes sense to define a $K_{(g)}$ as a means of comparison with other experiments. In general the significance of a K or K_g can be questioned when the distribution deviates strongly from the Gaussian. Oppositely, in cases where the distributions are not well known, but measurements of $(c_2)_{max}$ (we consider the most relevant case, p = 2) are available, one often assumes that the distributions are Gaussian which means that the relation $(c_2)_{max} = M (\pi\sigma^2)^{-1}$ holds, or $(c_2)_{\overline{max}} = (\pi\sigma^2)^{-1}$ if $(c_2)_{\overline{max}}$ is the maximum concentration per unit released. This combined with Equation 27 gives a K on the basis of $c_{max}(t)$. The resulting expression can also be taken as a new definition of a generalized K. In general terms it reads:

$$K_g \overset{def}{=\joinrel=} \frac{1}{4\pi} \frac{d}{dt} \left\{ (c_p)^{(1)}_{max} \right\}^{-\frac{2}{p}} \tag{28}$$

$$K_g \overset{def}{=\joinrel=} \frac{1}{4\pi} \frac{d}{dt} \left\{ (c_2)^{(1)}_{max} \right\}^{-1} \quad \text{for} \quad p = 2 \tag{29}$$

This definition is consistent with Equation 17 and identical with Equation 27 for Gaussian distributions. For other distributions it will deviate from Equation 27 (stronger as the distribution deviates more from the Gaussian), but it has the advantage that it always applies. Of course, it can only be used with measurements if a reasonable estimate of c_{max} as a function of time is available.

V. SHEAR EFFECT AND INSTANTANEOUS SOURCE

The following, rather different from the foregoing discussions, elucidates some of the observed particularities of the instantaneous source process. It can also be used for refinement of predictive models.

From the preceding section one may get the impression that only chaotic or at least eddy-like water motion can explain dispersion of patches. If all eddies are small compared to the patch, the classical diffusion concept applies; if the process exhibits an acceleration compared with the mode of growth predicted by an ordinary diffusion equation, it can be explained by a broad spectrum of eddies extending beyond patch sizes, the latter mainly referring to horizontal spread.

Several authors have discussed the combined effect of diffusion on a restricted scale and current shear in pipe flow,[72,73] surface waters, and atmosphere.[13,74] In connection with the preceding section, we will first consider analytical solutions for the case of an instantaneous point source. This case was treated extensively by Okubo[75,76] with a reference to Novikov.[77]

We return to Equation 6, but when trying to find analytical solutions for the instantaneous source, we drop the restriction that the mean velocity field be uniform. U, V,... = 0 can (and will) be maintained in the origin (where the release is made).

In his 1966 article[75] Okubo considers the horizontally two-dimensional case in a pure two-dimensional sense, i.e., neglecting possible vertical velocity imhomogeneities which can certainly affect horizontal spread, as remarked in the beginning of the preceding section. To keep the problem analytically tractable, Okubo restricts velocity variations to linear changes in space, so the (mean) velocity field is stationary with U and V depending upon x and y linearly. Further he requires nondivergence. This results in $u = \alpha x + (h-\eta)y$, $v = (h+\eta)x - \alpha y$, if

$$\text{(equal because of nondivergence)}$$

$$\alpha = \left(\frac{\partial u}{\partial x}\right)_o = -\left(\frac{\partial v}{\partial y}\right)_o \qquad \text{stretching deformation}$$

$$h = \frac{1}{2}\left\{\left(\frac{\partial v}{\partial x}\right)_o + \left(\frac{\partial u}{\partial y}\right)_o\right\} \qquad \text{shearing deformation} \qquad (30)$$

$$\eta = \frac{1}{2}\left\{\left(\frac{\partial v}{\partial x}\right)_o - \left(\frac{\partial u}{\partial y}\right)_o\right\} \qquad \text{vorticity}$$

Beside this "mean" velocity field a "small-scale" isotropic diffusion is assumed, in the classical sense (K constant), so that the transport equation simply reads:

$$\frac{\partial c}{\partial t} + u\frac{\partial c}{\partial x} + v\frac{\partial c}{\partial y} = K\left(\frac{\partial^2 c}{\partial x^2} + \frac{\partial^2 c}{\partial y^2}\right) \qquad (31)$$

Okubo finds a general solution for an instantaneous point source in analytical form for which the reader is referred to the original paper. One of the characteristics of the solution in this general form is that spatially it is a nonisotropic Gaussian distribution

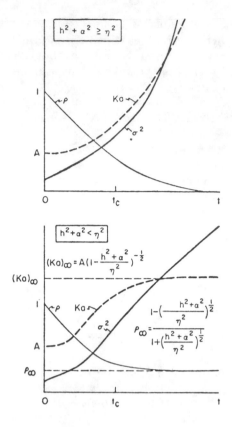

FIGURE 8.   Theoretical behavior in time of characteristics of two-dimensional diffusion in shear flow, after Okubo[75] (1966). $t_c = |h^2 + \alpha^2 - \nu^2|^{1/2}$, σ^2 = variance, K_a = apparent coefficient of diffusion, ϱ = ratio of the minor axis of the patch to the major axis.

with main axes making an angle θ with x and y. In Figure 1 of that paper, reproduced here as Figure 8, some important properties of the solutions are summarized. Some details look unrealistic (such as $\varrho \to 0$ for $t \to \infty$ if $h^2 + \alpha^2 \geqslant \eta^2$), but this should not surprise since of course in nature u or v will never increase indefinitely with x and or y. For a further discussion of the general cases $h^2 + \alpha^2 \geqslant$ and $< \eta^2$ we refer to the paper. Okubo observes, especially in the case $h^2 + \alpha^2 > \eta^2$, that the combined effect of the nonuniform convective motion and the small-scale diffusion accelerates the rate of dispersion to a great extent; if in the other case the extreme is reached that $h = \alpha = 0$, there is no acceleration at all (solid rotation).

Let us further restrict ourselves to the case of uniform shear, $\eta = 0$, $h = -\alpha > 0$. In that case, after Okubo,

$$(c_2)_{max} = \frac{Q}{4\pi Kt (1 + \frac{1}{12} h^2 t^2)^{1/2}} \tag{32}$$

For relatively large t ($t \gg 2\sqrt{3}/h = 4\sqrt{3}/\partial u/\partial y$), $c_{2max} \propto t^{-2}$, instead of $\propto t^{-1}$ as in the case of diffusion in a uniform two-dimensional velocity field. As said above, in practice the effect can only be temporary. It could be interpreted as the combined action of an eddy that is still relatively large compared with the patch and a field of turbulence with eddies small compared with patch size (e.g., generated by bottom friction). A little later the process could be repeated on a larger scale. The analysis gives not only some idea how the combination of small and large "turbulent" eddies can accelerate the dispersion, but also demonstrates that gradients in the "systematic" velocity field can speed up dispersion. We will return to this in the sections on grid approximations and on the distinct-particle approach (Sections VII and VIII).

Similar effects due to *vertical* velocity gradients have been considered by several authors,[76,78-81] also for atmospheric diffusion.[13,74] The combined effect of horizontal and vertical shear upon an instantaneous source in an infinite three-dimensional ocean has been discussed by Carter and Okubo.[82] It is interesting to note that again after an initial period with less influence of the shears, they find $(c_3)_{max} \propto t^{-2.5}$, a surprisingly small difference with Equation 32, considering that the number of dimensions as well as the number of shear "directions" were both increased by 1.

In a later paper[83] Okubo summarizes the earlier results and concludes that in a vertically bounded sea the time behavior of $(c_2)_{max}$ in the phase after the substance has spread over the vertical range between bottom and naviface is the same as in the purely two-dimensional case, i.e., $\propto t^{-2}$; the "small-scale" diffusion coefficient in the direction of the mean flow is only enlarged by the effective longitudinal coefficient due to vertical shear. In the sequel of the same paper, Okubo applies the idea of shear dispersion to the case of a continuous range of eddies in a somewhat simplifying way, by contributing the same characteristic velocity v* to the eddy (or eddies) delivering the main shear on a specific scale σ and to the certainly smaller eddies responsible for the (main) eddy diffusivity for the scale σ, while at the other hand he assumes that v* increases with scale as $\sigma^{1/3}$. Anyhow, the reasoning leads to the well-known result for an inertial subrange in two dimensions, discussed in the preceding section: $\sigma_{rc}^2 \propto t^3$ or $(c_2)_{max} \propto t^{-3}$. This result can be seen as an illustration of the fact that the *mechanism* of shear dispersion, sufficiently generalized, can explain the dispersion processes to a very large extent.

In a grid model (Section VII), the combined effect of the shears of the computed velocity field and (fixed-scale) diffusion coefficients can be regarded as an explicit computation of the shear diffusion (it would be better to talk about shear dispersion) in a very general sense. The question is only whether the computed field contains shear at the various scales in a sufficiently realistic way.

In a depth-averaged model, vertical shear and vertical diffusion do not occur. However, one can try to account for their combined effect upon horizontal dispersion in a parameterized way (Nihoul,[54] Section 3.5 and references therein).

The relative importance of shear dispersion caused by vertical shear and vertical diffusion can be estimated by simplified models with uniform horizontal velocities, expressed analytically (Kullenberg[2] and references, and Nihoul[54]). Nihoul has treated the (vertically) bounded case and concludes from comparison with experiments that during long periods (up to about 70 hr after an instantaneous release), dispersion due to vertical shear would be the main dispersive agent. Kullenberg considered the unbounded case, giving the limit of the time interval of its applicability ($K_z t/H^2 \leqslant 0.1$). His experiments (in areas with a small K_z) indicate that in the time interval mentioned, the horizontal spread due to vertical diffusion and shear is the main dispersive mechanism. As pointed out by Okubo,[14] in a vertically bounded sea the dispersion due to vertical shear and diffusion is only progressive in the initial stage and results in a constant contribution to the effective horizontal diffusivity K_h after the substance has spread homogeneously over the entire depth. This homogeneous phase is in fact the phase to which Nihouls' computations apply; consequently his $\bar{c}_{2\ max}$ decreases as t^{-1} and his effective dispersion coefficients v_1 and v_2 are constant. In shallow seas like the

Southern Bight or the North Sea, vertical spread occurs within a few hours up to about 1 day after release. However, thereafter, according to the experiments, the total K_h keeps increasing, approximately as $t^{1.0}$, instead of remaining constant for some time. This means that the horizontal dispersion due to vertical shear and vertical diffusion loses its possibly dominant role soon after vertical spread has taken place or earlier. This seems to be in contradiction with the relative long times (like 68 hr) mentioned by Nihoul,[54] for which the effect would still be dominant.

VI. SUPERPOSITION AND CONTINUOUS SOURCES

The principle of superposition is fully applicable for each separate contaminant as long as it is fully passive. It means that various concentrations of the same passive substance or property, at the same place and time, coming from different or identical sources can simply be added in order to obtain the total concentration at the point. As a special case (the most common application), the distributions due to a continuous (thus infinite) series of instantaneous releases from one source can be added, resulting in a distribution that corresponds to a continuous release. If the "instantaneous" patches are given analytically, the summation may indeed be tractable in continuous form, i.e., by integration. Otherwise the infinite series has to be replaced by a finite number of short successive injections to approximate the continuous case. In that instance the summands need not to be analytical functions; for example, they can be measured distributions.[84,85]

The idea of superposition, usually with the purpose to simulate a "plume", was applied and worked out by several authors[29,43,58,63,84-92] and can be found in various reference works and textbooks on surface waters as well as on the atmosphere.[93-95] In most cases the instantaneous elements or basic functions are expressed analytically, only Harremöes[84,85] seems to proceed straight from the measured distributions of instantaneous release trials. Okubo and Karweit include asymmetry caused by horizontal shear. In some cases the integrated results are presented in analytical form as well, but this is often restricted to the classical diffusion concept (constant K),[91,93,96,97] although not always.[58,89] Most authors consider a steady (mean) current field with uniform direction. This field is usually homogeneous, with a few exceptions (shear flow: Okubo and Karweit,[90] Csanady[13]). Analytical solutions always regard steady-state distributions. For nonsteady flow such as tidal currents and varying residual currents, numerical integration is necessary.

All operational applications for surface waters, as far as known, have been two-dimensional. For atmospheric pollution three-dimensional (analytical) superposition models are not uncommon.[95]

Rigid horizontal boundaries are either omitted (or neglected) or their influence is taken in account by simple reflection (mirror image of the source). This is usually restricted to one straight coastal bound. Van Dam[63] has constructed a model allowing for two parallel barriers, accounted for by repeated reflections, continued until the contributions from the remaining images are negligible. This can be useful for applications in estuaries and rivers. At present, the possibility for a second pair of parallel boundaries, perpendicular to the other one or two, is being built into the computational program.[98] This possibility can be used for approximations regarding bays and lakes closed at three or four sides. Of course these approximations remain very schematic.

All applications referred to have in common the fact that they do not allow for differences in water depth or layer thickness as a function of place. A gradual change in time, such as a slow uniform growth of the contaminated layer thickness of the individual patches by vertical dispersion, can be easily simulated in those models where integration is done numerically. No method has been reported so far dealing with decreasing depths near a coast. The lack of such a method is a serious drawback. A solution could be to distinguish between a number of layers,[99] say about five, each ending on a specific depth contour; these contours would have to remain straight lines. The restriction that closed boundaries have to be straight, is rather difficult to overcome. Figure 10 just illustrates an imperfect attempt to do so to a limited extent. Altogether it must be recognized that the method is limited in its applications, and that each further refinement, such as layering, should be weighed against switching to other modeling techniques (refer to following two sections).

Beside the rather widely spread application of superposition in models of atmospheric pollution, it seems that the most sophisticated superposition model for surface waters so far reported is the two-dimensional superposition model of Rijkswaterstaat, Netherlands.[63,92] This model has perhaps the largest history of operational use (1963 until now) in practical water pollution problems.[27,29,43,68,100-102] Some typical results are presented in Figures 9 to 11. Details of the numerical procedures can be found in the references.[92,63]

In general terms, the (volume) concentration in a point (x,y) at time t_m due to a release $Q(t)$ at (x_o, y_o) during the period $(0, t_m)$ (in which Q may be temporarily zero) is

$$C(x,y,t) = \int_o^{t_m} \frac{Q(t)}{h(t)} \bar{c}_2 \left\{ x - x_M(t_m,t), y - y_M(t_m,t), t_m - t \right\} dt \qquad (33)$$

in which $\bar{c}_2$ is the chosen function (on the basis of local measurements or theoretical considerations) describing the two-dimensional development of an instantaneous unit release (e.g., Equation 8, with $M = 1$); $h(t)$ is effective layer thickness, usually taken constant, but it can be used to account for initial vertical spread, e.g., from a top layer (caused, for example, by some buoyancy at the outset) gradually to full depth. $Q(t)/h(t)$ can be handled as a single function $Q_{eff}(t)$.

The development of an individual patch is represented by $\bar{c}_2 (\xi, \eta, \theta)$ if ξ and η are the coordinates with respect to the moving center x_M, y_M of the patch, and θ its "age". The expression for $\bar{c}_2$ could be made explicitly dependent on temporal influences such as tidal phase and weather conditions. In the present Rijkswaterstaat model, this has not yet been done because there is so far no empirical evidence for doing so, but the $\bar{c}_{2\,max}(t)$ function can be flexibly adapted[63] to characteristic curves for different areas and time periods and is not restricted to a simple form $c_o t^{-\beta}$ with constant β.

$[x_M(t_m,t), y_M(t_m,t)]$, for shortness: $x_{M_i}(t_m,t)$ in fact represent integrals of the form:

$$x_{M_i}(t_m,t) = x_{o_i} + \int_o^{t_m} U_i(\tau) \, d\tau \qquad (34)$$

where $U_i(\tau)$ is the velocity history of the center of a patch, leaving from (x_o, y_o) at time t. This is a particular kind of Lagrangian velocity; it is not the same function as the velocity history met by a single particle leaving from (x_o, y_o) at the same moment. Because of the growth of the patch, U_i should be a mean over a gradually increasing water volume, corresponding with the expanse of the patch. In other words $U_i(\tau)$ becomes smoother as a function of time when time proceeds and correspondingly x_{M_i} will become smoother as $t_m - t$ grows. Consequently, $x_{M_i}(t_m,t)$ viewed at as a curve in the x_i plane will become smoother as one moves from the source down the curve, the "center line" of the plume (Figure 12). This nuance will usually be neglected in prac-

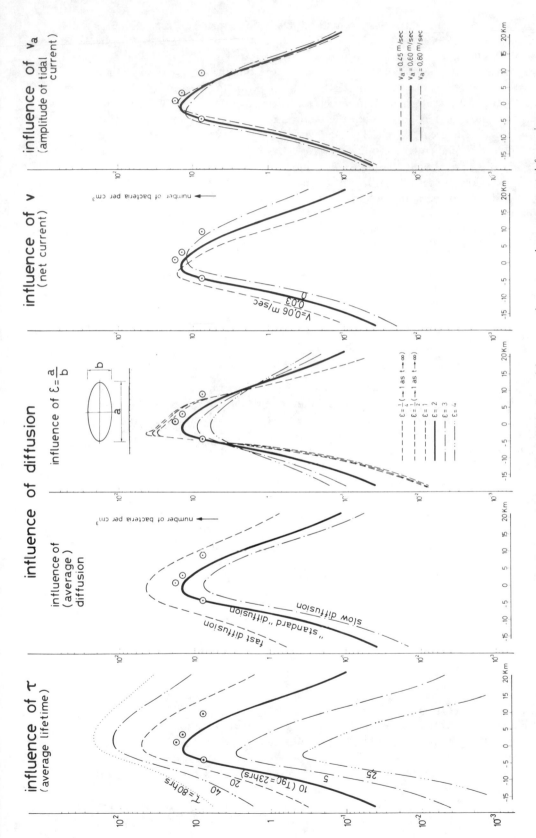

FIGURE 9.   One-dimensional output of superposition model. Coliform bacterial concentrations at sea shore as computed for various conditions.[102]

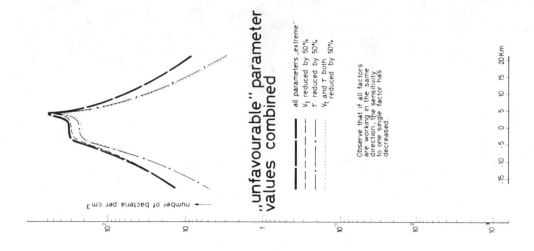

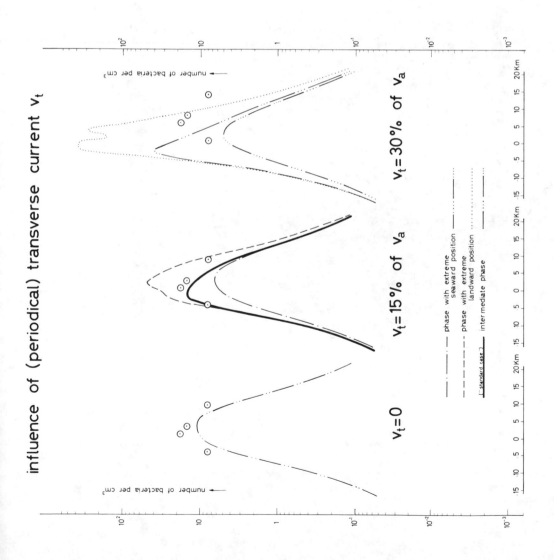

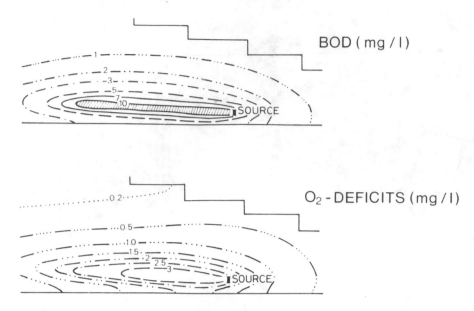

FIGURE 10.   Two-dimensional output of superposition model.[63] Example of a B.O.D./D.O. prediction with tidal flow in estuary. For each section (of constant width) a separate computation was made.

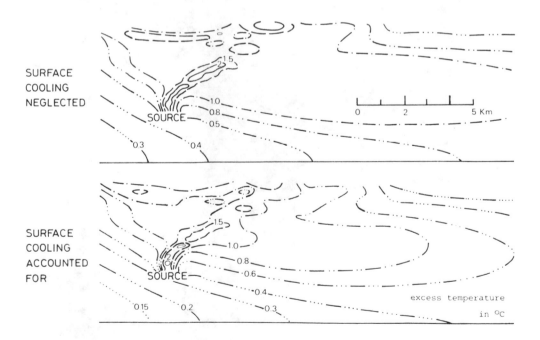

FIGURE 11.   Two-dimensional output of superposition model.[63] Example of predicted temperature distributions for two different conditions (surface cooling accounted for and neglected). Complicated flow pattern with tidal and residual currents.

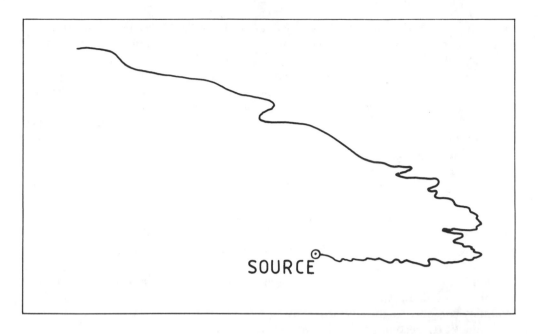

FIGURE 12.   (A) Sketch to illustrate effect upon center line of averaging over increasing area with increasing age and area of released material. (B) Aerial photograph suggesting a similar effect.

tice, also if x_{Mi} is determined on the basis of hydrodynamical computations in a grid model (Section VII).

In general, the two components x_M and y_M cannot be determined independently. Both components of the velocity at time τ will depend upon the position reached at that moment. The computation of the entire curve x_{Mi} (t_m, t) by a procedure coupled to a hydrodynamical computation is very laborious. For, say, 200 points of the curve, a 400 integrals of the type in Equation 34 have to be approximated. The number of time steps to approximate one integral will be of the order 100 as an average (number of steps increasing with $t_m - t$). To avoid this kind of computations, simplifications will often be made.

A great gain is obtained by assuming a homogeneous velocity field. Then the number of integrals can be drastically reduced. All patches released before any time t_1, between t and t_m, have different positions at t_1, but from that moment their displacements are all equal. This also means that the curve along which the patch centers are arranged at t_1 will not change in shape; it can be translated as well as be rotated. Of course, the assumption of a uniform velocity field is not realistic.

It is interesting to observe, though, that a homogeneous velocity function V_i (t) does exist, giving exactly the same function x_{M_i} (t_m, t) as the "real" field U_i (x, y, t) would do. This somewhat theoretical remark can be of use in specific cases if one has a computing program that only "knows" time-dependent velocity functions. For example, the distribution in Figure 7 has been obtained by adapting V_i (t) by trial and error in such a way that the initial part of the heat plume bends around an obstacle in the "correct" way. The limitations of the superposition method as regards rigid boundaries imply that the obstacle as such is not visible and some heat is also found where in the prototype there is no water at all. In more complicated cases such as systems of channels and shoals, the "approximation" of the real velocity field by means of a function V_i (t) becomes too difficult and too artificial. Besides, in those cases, the tractability of the superposition principle also becomes doubtful because of the complex geometry.

With a homogeneous velocity field U_i (t), the integrals of Equation 34 are not only drastically reduced in number because of the similarity of the various tracks, but they are also much easier to deal with because U_i (τ) does not depend implicitly on position and thus U_1 and U_2 are also mutually independent; they are just two functions of time. If they are given analytically, e.g., a sum of harmonic functions and constant or linear functions of time (representing combinations of tidal components and constant or slowly changing residual currents), they can often be integrated analytically and the resulting analytical functions x_{M_i} (t_m,t) can be substituted in Equation 33. However, even if Q or Q_{eff} is constant, the integral of Equation 33 cannot be found analytically as soon as x_{M_i} contains harmonic components which is a quite natural thing in the sea. So a general computation program for actual application of superposition in a marine environement will need numerical integration procedures for at least the calculation of Equation 33. However, only straightforward integration is required (i.e., simple addition of a series) so that no stability problems will arise. Only accuracy has to be watched. Especially at positions at or near the axis of the plume, the integrand exhibits strong, short periodic changes (pikes) if there is a tide, so that standard integration procedures may sometimes fail or work very uneconomically. For details of integration procedures, refer to the references.[63,92]

Further simplifications of the velocity field ($U_2 \equiv 0$; U_i = constant) may permit analytical expressions or approximations of C(x, y, t) according to Equation 33, dependent on the form of $\bar{c}_2$. For tidal seas they are only relevant for short term releases or for very large scales.

For the case $U_2 \equiv 0$, U_1 = constant, but $\bar{c}_2$ asymmetric due to shear of U_1, Okubo and Karweit[90] report no analytic solution but a result obtained by numerical integration (Figure 13). Analytic solutions (for certain distribution functions) can be obtained if U is constant and uniform as well.[58,89,91,93,96,97] These analytic solutions usually fail as soon as the contaminant is not conservative but breaks down with a characteristic time $\tau < \infty$. Sometimes it is possible to find approximate solutions in analytic form for special conditions such as relatively large U, in cases not allowing analytic solution otherwise. Brooks[103] derives such solutions, not by using the superposition integral of

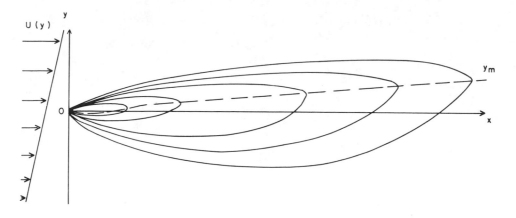

FIGURE 13.   Lines of equal concentration in continuous plume in shear flow after Okubo and Karweit[90] (1969).

Equation 33, but by approximate solution of the diffusion-advection equation for a continuous source.

The general principle of superposition, as formulated at the beginning of this section, further allows combination of various point sources with mutually different $Q(t)$ (also instantaneous if wanted), and simulation of a line source by approximating the line by a series of points. In the latter case, specific tricks can be worked out to keep computing times acceptable.[92] A result of such a computation is represented in Figure 14. However, an essential error, though it may be small, is made in such a computation as long as one function $U_i(t)$ or $V_i(t)$ (see foregoing) independent of position is used, so that the centers of all patches leaving at the same time from the various parts of the line source will move parallel to each other. In nature, however, they diverge in much the same way as a single patch of about the same "width" as the source length would grow. A better description is hard to obtain with this type of model. It should be added here that the observed error has some connection with errors that are introduced by using symmetric forms for the basic functions $\bar{c}_2$. The result obtained by Okubo and Karweit, giving an account for the influence of a current shear upon the shape of the individual patches and next upon the plume (skewness, Figure 13), illustrates that errors are made by neglecting these effects. However, accounting for it by means of the approximation of a uniform shear does not help; the real current field is much more complicated.

In the above, various limitations of superposition approximations have been mentioned:

1. It is impossible to deal with complicated boundaries.
2. Variations in (effective) water depth can only be accounted for by introducing a multilayer concept, which is not always adequate, especially with alternating depths.
3. The influence of the velocity field upon the internal structure of individual patches cannot be simulated, except in some special cases.
4. For similar reasons a systematic error is made in the simulation of a line source.

A point that was not mentioned yet can be added: except for some special cases (see Section X on decay and interactions), it is not possible to account for reactions between various substances.

There are alternative possibilities for modeling of dispersion that will be discussed

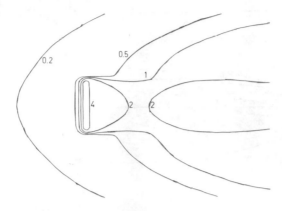

FIGURE 14.   Two-dimensional output of superposition model.[63,92] Line source in tidal flow. Directions of momentaneous flow and residual current are both to the right.

in the following sections. Of course there are advantages with the superposition approximation that make it often preferred in spite of its limitations and the existence of alternatives.

1.   Superposition models are relatively simple and relatively economic as regards computation efforts. In some cases semianalytic or analytic solutions can be applied. Even a rather sophisticated version of the model can be fed relatively easily with the required input and rather quick answers can be obtained. Because of these advantages the method may sometimes be preferred, also for predictions, and some inaccuracy be accepted.
2.   In general, a relatively simple model is suitable for preparative work and gaining insight. For example it can be used for sensitivity analysis (Section XI) in order to know the relative importance of various factors. The results, in turn, can be used as a guide for further experimental work (e.g., study of residual currents) as well as when constructing more complex models. Some of the studies mentioned above, such as on the influence of shear on a plume, can also be regarded as sensitivity studies.
3.   There are no computational problems in the sense that instabilities may occur or must be avoided or that unwanted numerical diffusion may cause errors.
4.   Decay can easily be accounted for; age distributions can be computed (contrary to finite difference models). For these two points refer to Section X.

VII. FINITE-DIFFERENCE AND FINITE-ELEMENT APPROXIMATION

Computing approximate solutions of the transport equations in discrete time steps on a fixed grid, rectangular or other, is one of the possible ways to profit by computed velocity fields on the same (or a closely related) grid. A second way to profit by such computed fields (but in a more restricted way), was briefly touched on in the preceding section: determining the centerline x_{M_i} (t_m,t) of a "plume" at time t_m (with t the "age" along the plume) by explicit computations of the positions x_{M_i} by integrals of U_i (τ) (Equation 34), deriving U_i (τ) step by step from the given field U_i (x, y, t). A third way to use U_i (x, y, t) will be discussed in the next section (distinct particle simulation). This third way does not have the essential restrictions of the superposition approach, nor some of the practical limitations of the present method (this section).

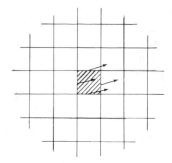

FIGURE 15. Instantaneous "point" release on rectangular grid.

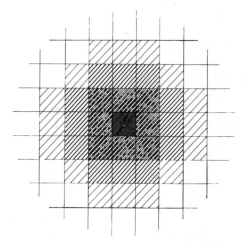

FIGURE 16. Instantaneous "release" with smoothed character as compared to Figure 15.

The importance of detailed knowledge of water velocities was already discussed in Section II. It was stated that if the knowledge is detailed enough, one needs in fact nothing else. Let us suppose for a moment that realistic knowledge of the velocity field, sufficiently detailed in space and time, is available. Then there are various possibilities to use the velocities for computation of transport of substance. Probably the most common technique is to assign one concentration to each "cell" (square or element) of the space grid and to account for convective transport by shifting at each time step some of the material in the direction of the local velocity. A diffusive or exchange transport between neighboring "cells" can be added, always bringing material from a place with higher to a place with lower concentration. Various computational schemes are possible to perform these operations.[4,104-106] Apart from the fact that no detail finer than that which corresponds to the grid size can be represented, systematic numerical errors are unavoidable. Especially, the convective transport, combined with a relatively small diffusion, may create problems. Let us consider an instantaneous point release in an interior point, probably the most tricky case if we take the "point" as small as we can: one cell, let us say a square (Figure 15). Often velocities are given on the sides of the square. The components perpendicular to the sides are the agents for the transport to and from the neighboring squares. A common way to estimate the concentration at the boundaries of the cell is to average between the cells at either side. Consequently, the inward convective transports in the center square in Figure 15 at the first time step result in negative concentrations in two of the neighbor squares. To keep all concentrations positive, a relatively large diffusion would be necessary. The outward transports at the other two sides cause a positive concentration in two other squares, uniform over the entire cells. The resulting situation is quite different from what we actually want: a small shift of the contaminated square as a whole, and maybe also some diffusion in all directions. The discrepancy becomes less striking when we deal with smoother distributions. An instantaneous release should start with a less pronounced distribution, such as the one shown in Figure 16. This is a limitation; we often want a somewhat finer resolution in the concentration field than corresponds to the grid of the velocity computation, rather than the opposite.

Negative concentrations and numerical diffusion can be suppressed by the use of special computation schemes.[104] None of these is completely satisfactory. Measures to suppress negative concentrations may increase numerical diffusion.[8] Finite difference schemes which are acceptable with an oscillating current may fail with a steady current.[8] According to Maier-Reimer it is only possible to meet all requirements (positive concentrations, conservation of mass, a sufficiently small numerical diffusion, and without sources, maximum concentrations which never increase) if the transport equation is solved in a Lagrangean grid (grid points move with the water mass). But always the fact remains that the obtained distributions are given on a relatively coarse grid; or, looking from the other side, sufficient refinement of the grid gives serious problems with respect to computing time and memory space. Maier-Reimer[8] remarks that usually (e.g., with point sources) the concentration varies much stronger in space than the velocity field. This suggests that when computation on a grid must be used (e.g., if reactions among different chemicals have to be included), the transport equation should be solved on a finer grid than the grid on which the equation of motion has been solved or the velocities are given. This solution does not sound attractive if this finer grid has to be a Lagrangean one, as Maier-Reimer has stated. In that case his conclusion that a distinct particle approach is to be preferred could be right. It should be said here, however, that refinement of the grid has a favorable effect with regard to negative concentrations and numerical diffusion, so that a Eulerian transport computation on the fine grid might be acceptable.

Positive results of computation of dispersion on a fixed grid have been reported by Leendertse[6] as mentioned in Section II. This concerns a bay with a narrow entrance, a pronounced bottom topography, and shoals which fall dry at low water; further, it regards short-term computations (a few tidal periods). No conclusions can be drawn for large or relatively even areas and long-term processes. In the latter case no serious problems are to be expected as long as we have widespread distributions (compared with grid size) and moderate gradients (no small sources). Good results may be expected with fresh water spread from well-mixed estuaries and, for example, spreading of radioactive waste or other materials in the North Sea entering through the channel.

Generally speaking, no details smaller than a few meshes of the grid should be expected. Problems with a small source will often require a local submodel with a finer mesh; if necessary this "nesting" of models can be repeated once more. It may be rather laborious, certainly if it concerns both water movement and transport computations. It may be acceptable in certain cases to use the finer mesh for the transports only, as suggested above.

If for long-term dispersion processes (several weeks up to about 1 year), water movements and transports would have to be calculated explicitly with constant resolution (corresponding to a constant grid and a constant time step), the extent of the necessary

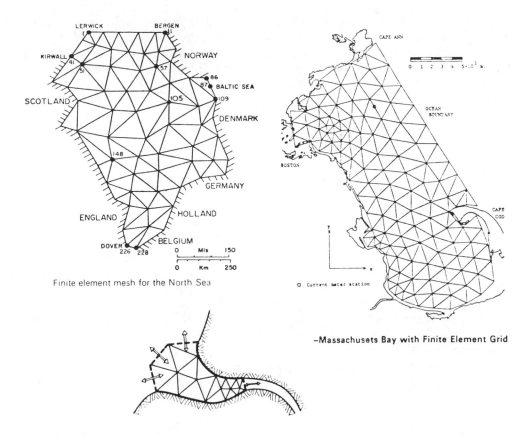

FIGURE 17.   Examples of finite element grids (North Sea,[111] Massachusets Bay,[112] river mouth[108]) illustrating adaption to coasts and local refinement of the mesh.

computations, with present computer technology, would be prohibitive. However, long-term spread corresponding to large scales, can be computed on a coarser grid than the short-term process (more recent events), corresponding to smaller scales. On the basis of this principle, Johnson and Leendertse[7] have designed a scheme in which a fine and a coarse grid are used successively in the entire model area. The fine grid is used for the recent processes (say, a few days), the coarse grid (squares 3 × 3 times larger) for the preceding period (weeks or months). With the coarse grid, correspondingly longer time steps can be taken, the computing time for the period computed on the coarse grid is reduced by almost a factor of hundred. No reports have so far been received on a complete program and operational use of the method. The general idea has some similarity with the increasing time step while going from present to past[63] in the Rijkswaterstaat superposition model.

The rectangular grid of a finite difference model has a limited capacity for adaption to coastline shape and orientation. Local refinement of the grid can only be established by nesting a second, possibly a third, model inside the first,[19,107] but a finite element[106,108-110] grid is more flexible. The orientation of the coastline can be followed and local refinement of the mesh is possible (Figure 17). Computations with finite elements are more laborious[105,113] than with finite differences (rectangular grid); the effects of negative concentrations and numerical diffusion are more or less the same.[108] The finite element method has become important rather recently,[114] when it no longer depended on a variation principle but was used in connection with methods of weighted residuals.[115]

A question to be answered when using a space grid in transport computations is which values are to be given to the diffusion or exchange coefficients. One often prefers the name dispersion coefficients, especially in less than three dimensions, in the latter case mainly to stress the fact that the exchange transport has components due to the combined effect of current shear and turbulent diffusion in the omitted space dimension (compare Section V). However, it is not only averaging or integration over left-out *dimensions* which influence the value of exchange coefficients. The finite size of the cells also plays a part. There is an analogy with the influence of patch size upon apparent or effective diffusion coefficients.[116] This implies that the *scale effect* is essential. If the grid size corresponds to the main scale or at least the largest scale on which notable exchange takes place, further increase of the grid size would not affect the coefficients. This however is a most theoretical case since it would mean that on the larger scales the velocity field would be homogeneous or at most represent "rigid" rotations and translations only. This means that in practice, the coefficients K in a two-dimensional grid representation will always increase with mesh size. From the standpoint that the differential equations and their finite difference approximation do not essentially change when space and time steps are varied, it may seem a paradox that one of the coefficients has to be adapted to the step size; such a thing certainly does not happen, for example, with friction coefficients in the equations of motion. The paradox is explained by the insight that when the mesh is enlarged so that velocities are averaged over larger areas, some information (inhomogeneities) is thrown away. The inhomogeneities have a dispersive effect that vanishes insofar as the inhomogeneities disappear. In a *global* way, this can be compensated for by increasing the exchange term by means of augmenting its coefficient. Another paradox seems to arise when we observe that the same hardly applies in the one-dimensional case. This is explained by the fact that in a one-dimensional approximation (e.g., of an estuary) the omission of two of three dimensions implies a complete averaging of velocities over these directions. All exchange effects thrown away by doing this have been put in the longitudinal mixing coefficient K. The remaining (extra) exchange caused by interaction between the exchange term $A(x) K(x) dC/dx$ and the relatively small longitudinal variations of the longitudinal velocities is so small that certainly its possible changes when the "mesh" dx is varied will be negligible as compared with the total exchange term.

The analogy between the exchange among cells and the exchange between a patch and its surroundings will help us to make a first estimate of K in a grid model. For a particular mesh dx in a (horizontally) two-dimensional grid model, K will be of the same order as the effective or apparent diffusion coefficient found at the corresponding scale[117] from a diagram such as shown in Figures 2B and 20. Preferably a diagram should be used based upon local measurements[116] or else upon observations from very similar regions.[117] This K roughly represents the exchange on the length scale equal to the mesh, including the (relatively small) contributions of exchange at smaller scales. In principle, exchange on larger scales will be taken care of by the computed velocity field combined with the exchange at scale dx. However, one can hardly expect that right from the mesh length and up, all details of the real velocity field are present in the simulation. There will always be some smoothing of the smallest details that theoretically could be represented on the given mesh. For this reason it should be expected that best results will be obtained with a K that corresponds to a patch that is somewhat larger than a cell of the grid. These considerations only hold as long as numerical diffusion is small compared to the applied K.

Definitive K-values can of course best be determined by verifying simulations on empirical data from experiments (artificial tracers) and measured distributions of natural tracers, such as salinity (river water content), temperature, and turbidity. It may occur that numerical diffusion is larger than or of the same order as required for the total exchange effect. In such a case no better results are obtainable than with $K_x = K_y = 0$ (two-dimensional case) for all t.

In the literature various proposals can be found[4,118,119] to relate to K to characteristics of the velocity field such as mean velocity, shear (see Section V), vorticity, and energy dissipation. Vorticity and horizontal shear can only be derived from motions on a somewhat larger scale than the mesh length. The larger-scale motions have relevance for the somewhat smaller scales to the extent they are "feeding" them. Dependence of K on characteristics of the currents should only be introduced as far as it can be demonstrated that it improves the correctness of simulations. In the three-dimensional case the vertical exchange coefficient K_z is strongly influenced by the vertical density gradient (Section IX).

VIII. DISTINCT-PARTICLE SIMULATIONS

Movement of distinct particles in the velocity field is exactly what dispersion is in nature. Passive contaminants are those which follow the velocity field and do not affect it. Particles (except extremely small ones) with a specific weight that deviates from that of the surrounding fluid so that they do *not* precisely follow the water motion, also invite us to treat them as distinct entities. In this section we will restrict ourselves to passive materials.

In Section II it was remarked that with sufficiently detailed knowledge of the velocity field, one could describe the dispersion process at the scales of interest. Strictly speaking there are a few exceptions like the phenomenon of salt fingering[2] which cannot be explained without explicit consideration of the role of molecular diffusion. Such are not the cases that we have in mind in the present section. One should rather think of shallow seas or coastal areas where we need models for prediction of pollutant spread in turbulent, vertically well-mixed waters although this does not necessarily mean that a distinct-particle approach would only apply under those conditions.

Our actual knowledge of the velocity field, for example, computed on a rectangular grid, is usually too limited for computing the dispersion of a number of particles realistically. Especially on the smallest scales of interest, the given velocity field will mostly be too smooth, even if we interpolate between grid points. To some extent this circumstance is analogous to the situation in finite-difference computation of dispersion where we need a diffusion or dispersion term and a coefficient K to make up for the "unknown" subgrid details of the velocity field. For particles, we can use a random velocity component for similar reasons. An important difference is that relevant subgrid details of particle distributions (concentration distributions) can be obtained (Figure 18 after Maier-Reimer[8]), resulting from the combined effect of the variation of the velocities in time, some subgrid variation of velocities in space (obtained by interpolation), and finally, the random velocity component. On larger scales, the more pronounced spatial differences in velocities corresponding to more distant grid points will enhance the dispersion and for particles sufficiently far apart, these differences between the "deterministic" velocities will become the main dispersing agent (provided the water motion simulation model is "good" enough) and the random movements could even be neglected. Therefore, the random effects should not be stressed too much by calling this type of model a "Monte Carlo" model. This name applies better for models in which no or little (deterministic) knowledge of the spatial variability of the velocity field is available so that proper simulation of dispersion must to a great extent or entirely be obtained by random motions. In that case, however, the same results can often be obtained much more economically by analytical solutions and superposition techniques (Sections IV and VI). Distinct-particle simulations are especially interesting where detailed velocity patterns are available, although there are some

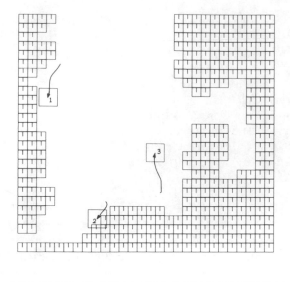

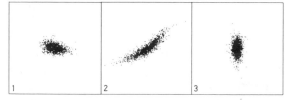

FIGURE 18 A

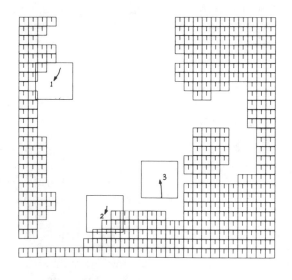

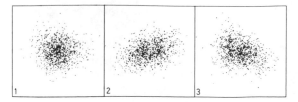

FIGURE 18B

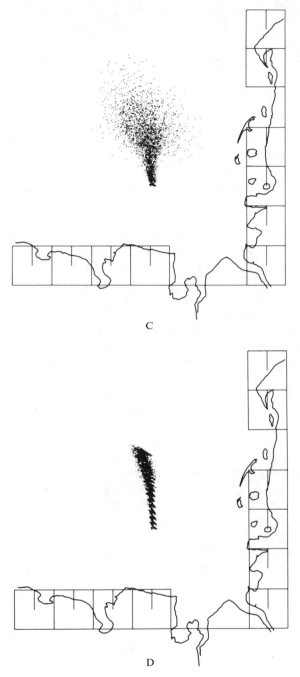

C

D

FIGURE 18. Examples of results obtained by Maier-Reimer[8] (1973) by his particle transport model (North Sea). (A) M2-tide, wind 28 m s^{-1} NW, (average) length of random displacement per time step constant in time, isotropic. Instantaneous release; 12 periods. (B) M2-tide, wind 14 m s^{-1} NW, (average) length of isotropic random displacement increasing as $t^{0.65}$, but additionally a random displacement in the direction of the momentaneous local velocity (as computed in the hydrodynamic model). (C) Continuous release, 20 M2-periods, wind 14 m s^{-1} NW, isotropic random displacement increasing as $t^{0.65}$; additional diffusion in direction of local velocities. German Bight. (D) Same but without isotropic diffusion, only diffusion in direction of flow.

other advantages. Depth differences and irregular horizontal boundaries can be taken into account rather easily, which is not the case with analytical solutions and superpositions thereof. In practice, the use and availability of more detailed velocity patterns and more complex representations of horizontal and vertical boundaries go very much together.

Actual computations of particle dispersion in the sea, based upon a velocity field computed on a grid (with the North Sea model after Hansen[120,121]) supplemented by random displacements have been performed by Maier-Reimer.[8,122,123] In 1978 a brief report was published[124] on a similar exercise performed on the basis of measured (and interpolated) velocities and the fluctuations thereof regarding an area in the western Black Sea. Thompson[125] reported particle simulations of dispersion in the atmosphere in 1971. See also Bugliarello et al.[126]

If at each time step of a computation each particle is given an independent random displacement, not depending upon place or time, a given set of particles (sufficient in number) disperses as if a constant diffusity is acting upon the "cloud".[127-129] This effect can be enhanced by inhomogeneities in the additional, given velocity field. In a computed velocity field the velocities within one or two squares of the grid are often rather homogeneous. Indeed it was found in the computations of Maier-Reimer that on this scale in most instances a cloud of particles did not disperse much faster than according to $\sigma \propto t^{1/2}$, although he also mentions a case in which the dispersion is obviously accelerated by a relatively strong local gradient.[8] In order to obtain a better agreement with experiments which indicate $\sigma \propto t^{\beta/2}$ with a β of 2.0 to 2.3, Maier-Reimer has applied the technique of gradually increasing the length of the random displacements proportional to t^a. He reports values $a = 0.50$ to 0.65. The moment all particles are in the same point is $t = 0$ or in the case of a continuous source, $t = 0$ is the time of release of the individual particle under consideration. It can easily be proven that an individual cloud (all particles released at $t = 0$) will grow according to $\sigma \propto t^{0.5+a}$ (in two dimensions) if the random displacements are the only dispersing agent. It also follows that the spatial distributions remain Gaussian. All this exists, provided the number of particles is sufficiently large. In other words, for instantaneous sources and uniform (deterministic) velocity fields, all solutions of the type in Equation 13 can be obtained in the described way, i.e., any $\beta \geqslant 1$ (two dimensions; more general: $\beta \geqslant p/2$) obtained by proper choice of $a \geqslant 0$.

One should realize, however, that actual computed velocity fields are inhomogeneous (as they should be) and inhomogeneities will become more important as a dispersing agent as particle distances increase. Therefore, it should be expected that for simulations up to a larger scale (relative to the grid size) than shown in Figure 18, a should gradually decrease with time after release and finally become zero, provided the velocity field is sufficiently realistic from a certain scale on. As remarked earlier, the remaining "constant" random displacements can finally be dropped as well, since their relative significance tends to zero if σ grows with a power of time larger than 0.5, due to the inhomogeneities of the velocity field.

The above solution to compensate for the insufficient detail of the velocity field in the initial phases of dispersion, as proposed by Maier-Reimer, is physically not satisfactory. Maybe one should not care too much about it if the concerned range is small compared to the entire range of interest, but usually this will not be the case. The physical objection is that in fact a particle never "knows" how "old" it is or when it was released. Only in the simulation we can "tell" the particle its age and consequently the step it should take. So it may happen that two groups of particles, accidently arrived at precisely the same spot at a certain moment (as may happen in the case of continuous release) will disperse quite differently, although in nature they would experience the same local influences. Although this is indeed unsatisfying from the viewpoint of physics, it may be that the influence of the incorrectness upon the (statistical) behavior of clouds and plumes is negligible. This may be doubted since the distributions obtained are all Gaussian (or superpositions thereof), while it is agreed (Section IV) that this is not the likely distribution in a field with eddies on all scales. Therefore, a more realistic simulation method would be preferable. This would ideally be a solution based upon a random velocity field[130] instead of random displacements. If we can represent such a field in a continuous way (preferably satisfying the continuity equation of the fluid), we would also meet the objection that the displacement method has the unrealistic feature that two particles in (almost) exactly equal positions take entirely uncorrelated random steps, just as if they would be far apart. At first sight a field consisting of a finite number of harmonic components (maybe five would be enough) of the form

$$\left[a_j \cos \left(\frac{2\pi y}{\epsilon_j \lambda_j} + \phi_{x_j} \right), \; \epsilon_j a_j \cos \left(\frac{2\pi x}{\lambda_j} + \phi_{y_j} \right) \right] \qquad (35)$$

$$(\epsilon_j = 1 \text{ in isotropic case})$$

possibly with some gradual temporal changes of parameters or ordinate origin to improve randomness, might do the job. The various values of λ_j ($j = 1,2,3, \ldots$) would simulate the scale effect up to the scale where the deterministic field "takes over". However, the spatial correlations must be correctly reflected in time as well. With strong systematic velocities such as tides, a particle moves so fast through a field like that of Equation 35 that it "forgets" local velocities too quickly. In nature, the relatively small eddies are carried along by larger scale movements. A first step to incorporate this effect would be to define field components like those of Equation 35 in a coordinate system moving at the speed of the systematic motion. A problem is that this systematic field is not homogeneous. It would be even more difficult to let the smaller eddies be carried along by the larger ones, but the first numerical experiments indicate that this is not necessary.[131] Studies of this kind are being undertaken at Rijkswaterstaat, Netherlands in the connection of North Sea transport models (Figure 19).

It should be emphasized that even the simple approximation of Maier-Reimer, modified as suggested earlier (decreasing α for larger t), results in a model that is much more powerful than the superposition method, and compared to a finite difference approach, it gives a much more detailed concentration field on the basis of the same velocity grid. The feasibility of the method has been demonstrated by Maier-Reimer et al.[8,132] and by recent studies in the Netherlands.[131,133,134]

In Figure 18 some of the results of Maier-Reimer are reproduced, illustrating the above statement. Maier-Reimer has used an extra random component in the direction of the mean flow (and opposite).[8] This can be seen as a way to account for the shear dispersion (Section V) due to vertical velocity shear. The analytical studies of Okubo and others (as referred to in Section V) can be used to estimate the value of the parameter concerned. One might consider a similar term for local wind effects (surface drift).

IX. THREE-DIMENSIONAL MODELING, VERTICAL TRANSPORT, AND DENSITY DIFFERENCES

The above three topics are closely connected. Vertical transports and density differences are not explicitly dealt with in horizontally two-dimensional models. Two-dimensional vertical models can be useful in estuaries and as a tool for numerical (or hy-

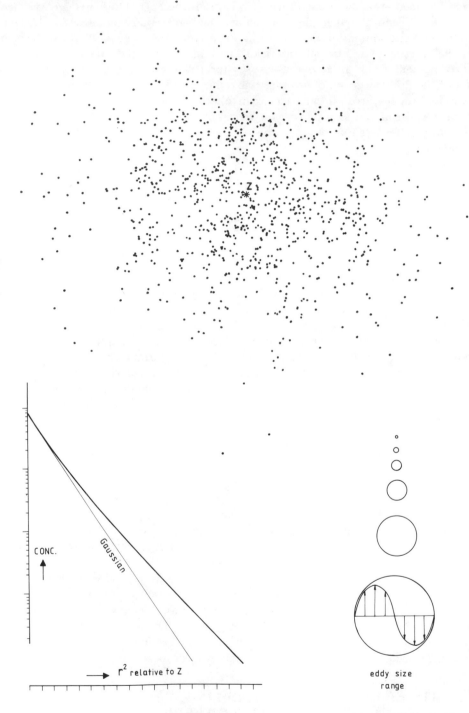

FIGURE 19.   Result of computation (288 time steps) of spread of 1000 particles in a field given by six terms of the form (Equation 35). Eddy sizes indicated on scale. Additional random steps are smaller than smallest eddies. Resulting distribution (heavy line) is non-Gaussian; Z = center of mass.

draulic) modeling of vertical mixing,[135,136] but not as direct means for description and prediction of dispersion in the (open) sea.

Modeling three-dimensionally does not necessarily imply the use of a three-dimensional model. As a first approximation, often sufficient, the vertical spread of substance can be treated separately. To illustrate this, we consider the simple solution of Equation 7 in three dimensions ($p = 3$), thus assuming constant diffusion coefficients and a homogeneous velocity field. Let the body of water be bounded by a horizontal plane at $\xi_3 = 0$ representing the naviface, while $W \equiv 0$. Thus the center of the distribution as well as the instantaneous injection at $t = 0$ are supposed to be at the surface $\xi_3 = \zeta = 0$. The sea surface is considered to be a perfect reflector which means that all concentrations are doubled as compared to Equation 7. At a fixed time t_o, the vertical distribution below an arbitrary point $(\xi_{1o}, \xi_{2o}, 0) = (\xi_o, \eta_o, 0)$ can be written

$$\bar{c}_3(\xi_o, \eta_o, \zeta, t_o) = \frac{2e^{-\frac{\zeta^2}{4t_o}}}{\sqrt{4\pi t_o K_\zeta}} \times \bar{c}_2(M, \xi_o, \eta_o, t_o) \qquad (36)$$
$$(\zeta \leq 0)$$

(for $\bar{c}_2$ see Equation 8). In other words, all vertical distributions are the same except for a factor independent of vertical position ζ; the effective penetration (e.g., expressed by a σ_ζ) is everywhere the same. This means that the distribution is identical to the one that is obtained if horizontal and vertical spread take place after each other instead of simultaneously. More generally, Equation 7 implies that the total distribution can be obtained by multiplying the distribution in one or two dimensions (any direction) by the relative distribution in the remaining direction(s). The relative distribution in one (two) direction(s) ξ_i is equal to the distribution of a unit mass in one (two) direction(s). Of all possible combinations the separation of Equation 36 is the most relevant for modeling of dispersion in the sea. Unfortunately, the underlying hypothesis of constant (relatively small-scale) diffusivities K_i in all directions does not hold well for marine dispersion (Sections III, IV, and V). We cannot even say that the separation of Equation 36 is sound as long as K_ζ ($= K_z$) is a constant parameter, independent of position and scale (which is often a fair approximation; see Chapter 1). The criterion for permissible separation of horizontal and vertical spread should rather be that the vertical velocity shear is not important compared to vertical diffusion in the sense that considerable asymmetry (in the direction of the mean current and opposite) and "inversion" of vertical distributions (concentrations increasing with depth at the upstream side of a cloud) do occur. It is clear that such a situation is in contradiction with the idea of similar vertical distribution at all places, as in the case of Equation 36. One may doubt the applicability of the separation in general when considering that in the case of strong vertical mixing where the shear effect is less important, one often does not need to describe vertical spreading at all, since the strong mixing will soon distribute the contaminant evenly over the entire depth. This doubt is justified to a great extent, but in practice one will often neglect the objection and use the separate treatment as an approximation of the average vertical penetration governed by an effective (not necessarily constant or uniform) coefficient K_z or K_v (compare Chapter 1). This concept does not depend upon the mechanism of horizontal dispersion so that it can be combined with forms other than Equation 8, such as Equation 13 (with any β), including other spatial distributions than the Gaussian, such as Equations 21, 22, 24, and 25. See for example the separate analysis of horizontal and vertical spread in the RHENO report.[32]

This all regards the instantaneous sources which in turn may be used as basic functions to obtain distributions of continuous sources by superposition (Section VI). This can be done fully three-dimensionally as in existing air pollution models,[95] but also in a limited sense as implied by the function h(t) in Equation 33 which in fact represents the effective depth of penetration, implicitly supposed to be the same for each element of the "continuous" series of patches. In this application the vertical distribution function as such is not considered; the concentration represents a surface value or an average over the contaminated layer (from surface to effective depth of penetration).

More precise descriptions of three-dimensional distributions can generally not be obtained by analytical solutions and superposition thereof. As an exception (for instantaneous release only) we can mention the analytical solutions for certain idealized cases of shear diffusion as referred to in Section V.[82]

For more precise three-dimensional modeling one must call in the numerical grid methods. In (horizontally) two-dimensional grid models for transport computations, three-dimensional aspects of mixing are accounted for in parameterized form. This especially refers to the effect of vertical diffusion and vertical velocity shear upon horizontal dispersion (Section V). A first step to make this mechanism more explicit is letting its contribution to the diffusivity tensor depend upon the momentaneous size and direction of the mean velocity.[8] The next step is introducing the vertical dimension explicitly, but combining the numerical description in horizontal sense with analytical approximations in the vertical direction.[137] One can restrict this to the concentration distribution by using the simple model with constant (effective) K_z. The following step can be the introduction of an analytical approximation of the vertical velocity distribution and computing shear diffusion explicitly.

If vertical and horizontal density differences occur, the only way for three-dimensional treatment of dispersion probably is the numerical computation on a three-dimensional grid. Transport computations of salt, heat, and pollutants are only possible in such a model if the water motion is also computed on the grid, and conversely the flow computation is only possible when combined with the computation of the changing density field. Because of this mutual dependence, flow and transport computations have to be performed simultaneously. Separate computation of the transport of passive pollutants seems at least rather unpractical, since vertical diffusion coefficients are influenced by the density distribution.

A computer program dealing with this most general case of water flow and transport of constituents has been developed by Leendertse et al.[138,139] Test runs have been performed for Chesapeake Bay and San Francisco Bay;[139] other tests concern inland waters. Recently a report was published on simulations of Bristol Bay[140] with this model, including salinity and temperature distributions.

Useful marine applications of a three-dimensional model of this kind are not only the computation of salt intrusion in complex estuarine systems but also pollutant transport in estuaries and in coastal seas with density gradients. The most involved applications are probably those where relatively large quantities of waste waters which induce artificial density effects are discharged in marine areas with natural density differences. It is clear that (with the exception of physical models; Section XIII) these complex situations can only be dealt with by three-dimensional grid models. Although at present these models, certainly with inhomogeneous density, are still in the phase of development and adjustment, they may become feasible and reliable tools in the near future. The current progress in computer technology is an important factor in this respect. On the other hand, it may be instructive to mention the most remarkable simulation of the distributions of oxygen, phosphate, and nitrogen in the Atlantic Ocean made as early as 1951 by Riley[141] on the basis of a finite difference form of the diffusion equation in three dimensions without the aid of a computer. In this connection we also mention the rather global estimates of long-term spread of radionuclides from dumpings, etc. for the Atlantic or the entire world ocean, on the basis of simple formulas such as Equations 36 and 13 and superpositions thereof.[91,142-145]

Important parameters in three-dimensional modeling are the coefficients of vertical exchange of mass and momentum. A discussion of observed values, mutual relations, and the influence of vertical density gradients has been given in Chapter 1. In (near) neutral conditions the coefficients for vertical transfer of mass and momentum are almost the same. The theory of the influence of (pronounced) vertical density gradients, which have a strong reducing influence upon all vertical exchange, has not made much progress during the past decennia.[2,146,147] The influence is usually expressed in terms of Richardson numbers but the relations have an empirical character. For modeling purposes the empirical approach is acceptable. The model itself can be used as a tool to find out which relations give the best fit to empirical data. Measurements in the prototype are often difficult and the environment conditions in field experiments can be rather complex and cannot be controlled. A mathematical model combined with laboratory experiments is a good tool for investigating this particular problem. For this purpose, the model needs not to be three-dimensional; a simplified, vertically two-dimensional model can be used, combined with a laboratory flume.[136]

X. DECAY AND INTERACTIONS

Decay and interactions of substances, properties, and organisms are not transport processes and from this point of view they do not belong in a chapter on dispersion. The processes themselves will therefore not be discussed here. They need, however, some attention in the context of dispersion models for the following reasons:

1. In general they cannot be separated from the transports in actual calculations; e.g., by taking them into account after completing the transport computation. An exception is simple decay in the case of instantaneous release.
2. The way they are incorporated in the models depends upon the type of model.
3. Sometimes they cannot be incorporated at all. For example, superposition models are restricted to simple decay and to a special case of interaction with two components (B.O.D./D.O.) and possibly another simple system of the same character.

The most simple and most common case of decay is the one that can be described by a constant half-life time $t_{1/2}$ or a constant characteristic time τ by which the amount of matter (etc.) decreases as

$$M(t) = M(t_o)e^{-\frac{t - t_o}{\tau}} \tag{37}$$

This means that the (relative) decay rate is independent of time. Equation 37 is correct for any $t_o < t$; it is not necessary to know when the substance was released or generated. This means that in, for example, a finite difference computation the total amount (of the same substance) in each "box" can be adjusted for decay at once, in each time step. In a short time Δt the fractional loss is $\Delta t/\tau$ which is the first order finite difference approximation of Equation 37 as a solution of the differential equation

$$\frac{dM}{dt} = -kM \tag{38}$$

in which the constant k evidently equals $1/\tau$.

A step by step solution of Equation 38 becomes a necessity if k depends upon local and/or temporal conditions. An example is the loss of heat, which takes place at the surface so that the loss per second is not determined by the total amount of heat (surplus) M in a water column but by the (surplus) heat concentration (excess temperature) at the surface, which is ($\propto$) M/h if h is the layer thickness or an effective penetration depth. If h is independent of time and position, the form of Equation 38 is not affected (h can be incorporated in k) and a solution of the form of Equation 37 is valid, with

$$\tau = \frac{\rho c_w h}{A_w} \tag{39}$$

(ϱ = density, c_w = specific heat, A_w = surface heat loss coefficient (Joules per m² and per °C excess temperature)). If h is place and/or time dependent, Equation 37 cannot be used and Equation 38 has to be solved stepwise.

A somewhat similar case occurs if neither Equation 37 nor 38 applies because the rate of decay of a contaminant depends explicitly on its age, which is sometimes the case with organic matter. Certain processes may only start after certain parts of the molecules have been oxidized first so that Equation 37 must be replaced by an equation

$$M(t) = M(0) \ (1 - F(t) \tag{40}$$

in which $0 \leqslant F(t) < 1$; $F(t) = 0$ at $t = 0$; $F(t) \rightarrow 1$ for $t \rightarrow \infty$ while $t = 0$ is not an arbitrary time but the time of injection of the particular fraction M(0) (the time it enters the environment where the decay process takes place). An equivalent of Equation 38 now does not exist which means that in a finite-difference model the step by step solution can in fact not be realized, except for the case that all the substance has been released at one moment. In a finite-difference model the various fractions of different age are not followed. In this respect superposition models are superior; the integrand of Equation 33 is just multiplied by $\{1 - F(t_m - t)\}$. In a finite-difference model such cases can only be treated by making the reactions more explicit, introducing the various intermediate products of the entire breakdown process as separate components.

Systems of several interacting components from various sources can only be dealt with by finite-difference models. Again under the condition that all reactions and intermediate products are made explicit, since these models can discriminate constituents only by identity and not by age.

Superposition models work essentially by age, building up concentrations at each point by adding up the contributions of the individual releases at the successive specified points of time in the past. This makes it easy to determine average age and age distribution of a substance or a set of organisms, in any point of the field, which improves our understanding in certain applications. This form of insight cannot be obtained by finite difference simulations.

A specific case of two components that can be treated by supersition is the case of organic matter and corresponding oxygen deficits. This is possible because there is only one substance which is released and the consequent oxygen deficit distribution is strongly coupled to that of the pollutant. The problem has been solved as soon as we can describe the distributions in space and time for both pollutant and oxygen deficit. It is assumed that oxygen concentrations do not become so low that the oxidation rate is influenced. Let us further assume that the breakdown or organic matter can be approximated by Equation 37. This is not a necessary requirement but in this case we obtain simple analytical expressions while otherwise a step by step computation must be used. The behavior of the total oxygen deficit $M_D(t)$ corresponding to the pollutant decay (Equation 37) is then given by

$$M_D(t) = \frac{\beta_r M_o}{\dfrac{\alpha_r}{h} - \dfrac{1}{\tau}} \left\{ e^{-\dfrac{t - t_o}{\tau}} - e^{-\dfrac{\alpha_r}{h}(t - t_o)} \right\} \qquad (41a)$$

$$\text{if } \frac{\alpha_r}{h} \neq \frac{1}{\tau}$$

$$M_D(t) = \beta_r M_o (t - t_o) e^{-\dfrac{\alpha_r}{h}(t - t_o)} \qquad \text{if } \frac{\alpha_r}{h} = \frac{1}{\tau} \quad (41b)$$

$M_o = M(t_o)$; α_r = reaeration coefficient (m s^{-1}); β_r = required amount of oxygen per second and per unit of pollutant. *Only* if quantities are expressed in "total B.O.D.", $\beta_r = 1/\tau$. The solution is continuous at $\alpha_r/h = 1/\tau$ so that Equation 41b has no special physical meaning; for practical purposes it can usually be omitted. Depth or layer thickness h occurs for the same reasons as it does in the heat loss equation; heat loss and reaeration both take place at the surface and become less effective (larger τ) as h increases.

According to Equation 41 the initial deficit is zero as well as the final deficit (t = ∞), as it should be. The spatial distribution of the deficit (for an instantaneous release) is equal to that of the pollutant, whatever it may be. However, for a continuous release, the superimposed distribution of deficits will not have the same appearance as the (superimposed) pollutant distribution (Figure 10). This is due to the difference in time behavior between Equations 37 and 41. The initial deficit being zero, a relatively high maximum deficit and large gradients near the source are absent.

Distinct particle models seem rather inefficient for simulation of reactions, except for simple decay or a system such as B.O.D./D.O., just briefly described. The problem is that for interactions between various components (with mutual independent distributions in space), concentrations have to be computed (by counting the particles per square) at every time step, and after applying the reaction matrix, the weight of the various kinds of particles in each square have to be adjusted accordingly.

XI. OPERATIONAL USE OF MODELS AND SENSITIVITY ANALYSIS

In this section we will briefly consider the use of dispersion models in pollution problems of current and local interest. The examples are restricted to two-dimensional modeling. Two main cases can be distinguished: the models are used for obtaining a better insight in a present situation of pollution (including calamities) in order to design measures or policies, or models are used as a predictive tool with the design of outfall works or other equipment in order to keep a future pollution within prescribed bounds. In all cases, the models need some input of parameter values and because of the complexity of phenomena and the limited scope of available theory, some minimum knowledge derived from local observation is usually necessary for obtaining some (minimum) reliability of the parameter values to be used.

Sensitivity analysis, in this context, could be defined as any effort to find out, by calculation or experiment, how accurate a parameter has to be known to obtain the required accuracy of prediction. We might generalize this concept on the basis of the following considerations. Required accuracies may be unrealistic in the sense that they do not allow for the natural variability of the conditions that determine pollutant distributions. An ideal prediction includes the fluctuations around the mean concentration. Models which generate these fluctuations explicitly are usually not available. The models described or mentioned in the preceding sections are largely based upon averages (like the ensemble average $\bar{c}$ of the instantaneous release, used when calculating

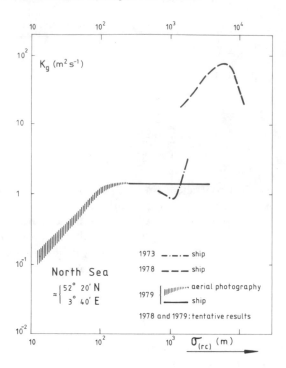

FIGURE 20.   Variability of patch diffusion at indicated position (tentative results). Note large difference between 1978 and 1979 experiments at scales of about 1000 to 3000 m. One should realize that in one experiment, the various scales are reached successively in time so that the corresponding K_g are not measured simultaneously. Definition of K_g as in Equation 29.

the integrated concentration C (formula of Equation 33) or will usually deal with one particular situation at a time (e.g., a particular set of boundary conditions used when computing the velocity field on a space grid). In principle, grid models have the possibility to include variability in a single run (of long duration) by varying tidal and meteorological boundary conditions on the basis of observed statistics. Using simpler models such as superposition models, it seems more appropriate to repeat runs while changing the various parameters one by one[27,102,148] and thereafter in a limited number of combinations[102] ("favorable" and "unfavorable"). This approach is feasible because of the relatively short computing times of these types of models. This especially applies if not a whole field of concentrations is computed but a limited part like a shoreline distribution. An example of such an approach has been given by Van Dam[102] (Figure 9), who remarks that the predictions were later compared with long-term observations and that a good agreement was found in terms of extreme values (favorable and unfavorable situations). On the other hand, this approach does not provide the frequency distributions of the various deviations from the average concentrations.[87,149] Csanady[13] in Chapter VII of his book gives an analysis of the variability problem from the viewpoint of turbulence theory. In the practice of marine pollution, the strong influence of transient but often semipermanent meteorological conditions seems to play a major role, at least at certain places. Repeated experiments during several years at precisely the same location in the North Sea have revealed an astonishing variability,[150] e.g., in terms of the effective overall dispersion coefficient K_g (Figure 20). The corresponding graph of $c_{2\ max}$ shows a divergence which goes beyond the bounds of the set of curves published earlier (Figure 3). Meteorological influences must be the cause of this great variability; maybe not only winds but also radiation, causing weak stratification. The various conditions, according to the $c_{2\ max}$ curves, last long enough to influence the dispersion process markedly for one to several weeks, possibly longer. The effects are so severe that they could have a considerable influence on the design of a possible artificial island for industrial purposes in such an area.[151]

As an example of sensitivity analysis, in Figure 21 the influence upon distributions of conservative matter, computed with a superposition model,[63] is given for the following parameters.

A. Diffusivity — In terms of overall effective [patch] "diffusion" coefficient K_g, on the basis of the two bounds ("extreme" cases) which span the North Sea data up to 1978 (Figure 3)

B. Residual current — Only in the sense that the assumed average value of 4 cm s^{-1} has been reduced to zero for limited periods (2 and 4 weeks). A considerably longer period of zero average net flow would give a more dramatic effect.[151] Its simulation would also require a somewhat more careful application of the superposition method, accounting for the random motion of a patch, which still occurs if the *average* residual flow is zero. The figure illustrates the great importance of residual currents for steady-state concentrations of long-living substances. This is one of the reasons of interest in residual current models from the viewpoint of dispersion. This regards currents induced by wind[122,123,152] and pressure fields[152] as well as other mechanisms.[123,153-155]

C. Distribution function — This "parameter" only applies for the superposition method, but the figure illustrates that if this method is used, the proper choice of the distribution function can be important. This would not be so apparent if the residual current was smaller. In the present case however there is such a pronounced difference in the "upstream" part of the field, that from natural data such as river water distributions, it may be possible to come to a decisive answer, at least about the *character* of the distribution function for the larger scales.

The given examples refer to a continuous release at the same distance from the coast of the Netherlands as where the above-mentioned extreme variability was found in recent years. With a conservative contaminant, as in the chosen examples, gradients are generally small, at least in the "downstream" area, and the concentrations are relatively unsensitive to parameters such as (patch) diffusivity and distribution function (except for the low concentrations in the "upstream" area). As observed, the general picture is strongly determined by the magnitude of the (average) residual current, as illustrated in this example of sensitivity analysis. This can also be understood by a more global approach. Long-term and large-scale averages of residual flow of a through-going (not rotating) character[156] have a determining influence upon *residence times*[123,157,158] and *flushing times*,[155] and thus upon average concentrations of conservative substances over the entire area. The importance of residual flow becomes rapidly smaller with decreasing lifetime of contaminants.[151] Conversely, the sensitivity to distribution function and diffusivity increases when the lifetime τ decreases. The same is true for the impact of transverse currents, at least for sources near to the coast.[102]

Sensitivity analysis can, sometimes for reasons of economics or availability, be performed with models simpler than the one finally to be used. This may even be done before a definite model is chosen, developed, or implemented.

Operational use of the various types of nonphysical models has already been mentioned in the preceding sections, especially in the connection of superposition models[27,43,58,68,85,91,101,102,151] and two- and three-dimensional grid models.[6,140] For operational use of physical models, refer to Section XIII.

It is not always possible to distinguish between sensitivity analysis and operational

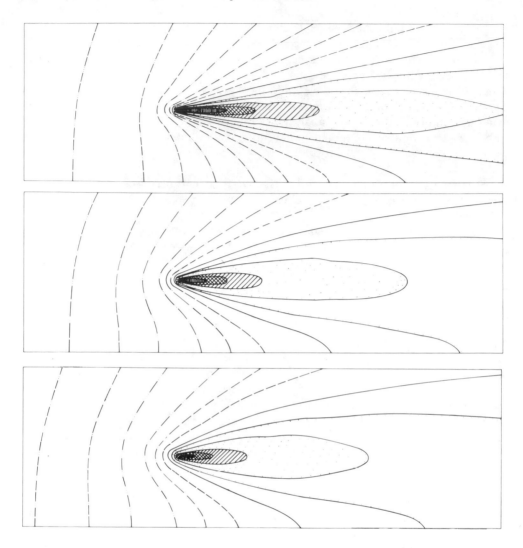

FIGURE 21.   Examples of sensitivity analysis with superpostion model[63]. (A) Diffusivity. For upper, middle, and lower drawing, results (conservative substance, continuous release during infinite time with 0.04 m s^{-1} residual current) correspond to Figure 3, curves S, M, and F respectively. (B) Residual flow. Current has been permanently 0.04 m s^{-1} (upper), 0.04 m s^{-1} but zero during the last 2 weeks (middle), 0.04 m s^{-1} but zero during last 4 weeks, respectively. (C) Distribution function. Upper drawing: distribution (Equation 24); lower: Gaussian distribution taken for individual "patches" (terms of superposition). For further conditions refer to Figure 21A (dilution process M).

use. For example, the computations made by a simple superposition model during the design phase of the Hague outfall[27,100] were mainly used in a relative sense: the results indicated that from a certain outfall length an appreciable improvement could only be obtained by a considerable and relatively expensive lengthtening. If further improvement would turn out to be necessary in the future, additional purification would be a more economical solution.

Models can only have practical value when they have an empirical basis. This usually means that they have to be verified or calibrated.[1] There is often a feedback[1] in the sense that first results and comparison with various data may lead to modifications of the model. Generally speaking, there is an interaction between modeling and observation.

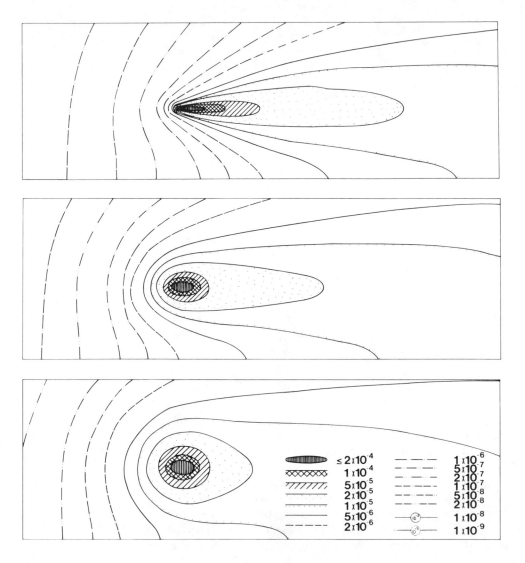

FIGURE 21B

Measurements for obtaining empirical information on model parameters may concern the velocity field as well as dispersion of tracers. The latter can be artifical[28,31,32,82,85,159] (dyes, radioactivity) or natural[35,141,160] (fresh water and salt, nutrients, turbidity, and heat). Experimental techniques will be treated in Chapter 3. Artificial tracers are often the only possibility to investigate the smaller scales since natural constituents usually do not come from small sources and the location of the natural source cannot be arbitrarily chosen. For the large scales, natural tracers may be quite suitable, especially river water which is a most conservative substance. Evaporation is usually negligible. Since it concerns the large scales, it is acceptable that the location of the source cannot be precisely chosen. A certain area around the river mouth has to be excluded, not only because it is difficult to define shape and size of the source but in most cases also because of the fact that the river water does not behave as a passive contaminant because of its large quantities, influencing current patterns, usually combined with induced stratifications.

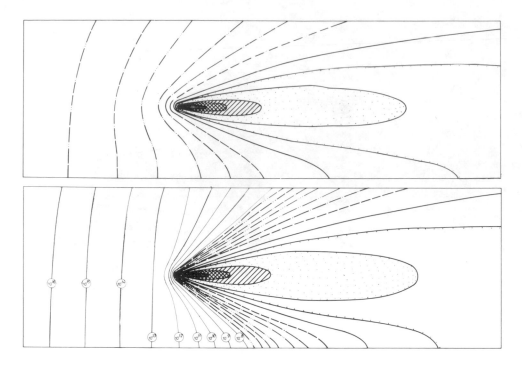

FIGURE 21C

Very interesting results have recently been obtained from the analysis of river water distribution along the Dutch coast.[160] During a period of several years, river water distribution had been monitored regularly. Thereafter the entire period was simulated by the superposition model. The residual current was adjusted until a "best-fit" to the data was obtained. The resulting large-space and large-time average differs significantly from the estimates based upon the measurements at fixed points in the past. The example shows how a model can be used as a tool in obtaining better input data for its operational use, the residual flow being an important parameter in operational applications, as shown by the sensitivity analysis described earlier.

XII. NONPASSIVE CONTAMINANTS

In the preceding sections it was mainly spoken of passive contaminants. One exception, the injection of large quantities of fresh water was only briefly touched in Sections IX and XI.

Four main types of nonpassive contaminants can roughly be distinguished: (1) fluids mixing well with sea water but introduced in such quantities that they influence the flow, at least locally; (2) fluids mixing well with sea water, but with a different density and introduced in such quantities that density effects influence their behavior, at least initially; (3) fluids not mixing with sea water; and (4) solids with a density different than that of sea water.

In most cases, fluids of the first category belong to the second category as well and vice versa. In practice it always concerns water masses, but different from the receiving basin in either salt content, load of other solvents, temperature, or a combination of two or three of these factors, which usually means that there is a density difference with the receiving water. The most important examples are industrial and domestic waste water, cooling water (either fresh, salt, or brackish), discharges of rivers, lakes, channels, etc. In the limited scope of this chapter we refer to what was already said about this category in Section IX. For materials of categories 3 and 4 that float or mainly move near the bottom we refer to the chapters on oil and on sediments.

What remains are the suspended particles, small organisms (bacteria), and suspended fluid drops. If the particles, etc. are small enough, the tendency they exhibit to float or sink is, at least in shallow seas, entirely compensated by the dispersive action of turbulence. In that case they are passive materials. What finally remains are those particles or fluid drops that are just large enough to be not (or not always) evenly distributed in the water column. Of course, the most important example are the fine sediments of intermediate grain size. In the context of this chapter we can restrict ourselves to the mechanism on which the vertical distribution depends.

The simplest model of the balance between turbulence and falling rate (positive or negative) is based on the assumption of a vertical diffusion coefficient K_z that is independent of depth. In a stationary current it will be constant in time as well. For the equilibrium state we can write (w_p = vertical particle velocity in still water [falling rate])

$$w_pC - K_z \frac{dC}{dz} = 0 \tag{42}$$

If the suspended mass in a water column of 1 m² and height H is M kg, the solution reads ($z = 0$ at the bottom and positive upwards):

$$C(z) = C(0) \, e^{\frac{w_p z}{K_z}} \tag{43}$$

where

$$C(0) = \frac{w_p M}{K_z \left(e^{\frac{Hw_p}{K_z}} - 1 \right)} \tag{44}$$

In this form the formulas are the most convenient in the case $w_p > 0$ ("light particles" such as tiny oil drops), but if $w_p < 0$ ("heavy particles" such as fine sediments) the formulas are more lucid when we write:

$$C(z) = C(0) e^{-\frac{|w_p| z}{K_z}} \qquad (w_p < 0) \tag{45}$$

where

$$C(0) = \frac{|w_p| M}{K_z \left(1 - e^{-\frac{H|w_p|}{K_z}} \right)} \qquad (w_p < 0) \tag{46}$$

For a particular substance, let us say fine inorganic sediment, the solution represents a different distribution for each size fraction since w_p depends upon size according to

Stokes' formula. The formulas do not require sedimentation if K_z diminishes (e.g., during a tidal cycle), but just a redistribution. This is indeed what seems to happen in many coastal seas where some periodic change in vertical distribution of fine sediments during a tidal cycle is observed,[161] but the same materials are not found sedimented on top of the hard sandy bottom. The changes in distribution are often moderate, which provides a means to estimate K_z in such areas. A typical value of $|w_p|$ is 1 m/hr. This combined with an intermediate value of C(bottom)/C(surface) = 2 leads to $K_z = H/2500$ m^{2}s^{-1}, e.g., 5×10^{-3} m^{2}s^{-1} if H = 12.5 m. The almost uniform distributions often observed indicate that even higher K_z values occur temporarily. One should realize that in fact there is no equilibrium state as required by the formulas. The sediments need time to redistribute and the tidal period is very short compared to the order of magnitude of w_p and water depths while turbulence needs time to die off. These two effects will usually cause a great damping of the periodic distribution changes compared to equilibrium states (which are not reached) and also a considerable time lag,[161] e.g., between maximum velocity and maximum concentrations in the top layer.

In a vertically one-dimensional finite-difference model the process can be simulated dynamically, including depth dependence of K_z related to the vertical velocity distribution. Such a model can in turn be included in a three-dimensional model (Section IX), dealing with various fractions separately, if desired. In that case source and sink terms should be added, taking care of erosion and sedimentation, controlled by a criterion to be chosen from the pertinent literature.

Schubel et al.,[162] in relation to field investigations of turbidity generated by open-water pipeline disposal operations, have used a vertically integrated suspended solids plume model with the assumption of an effective (net) settling velocity which results in an ordinary first-order (exponential) decay. Their model is further based upon the Okubo-Pritchard[37,57,58] relationship for horizontal diffusion (our formula (Equation 12) for p = 2, β = 2). In a number of cases the model of Schubel et al. agrees reasonably well with observations.

Others have used a two-dimensional model for the spread of dumped dredged material with neglect of settling effects (Terwindt[163]). Measured surface concentrations in the concerned area are lower than predicted, indicating an effect of settling although it has been shown that this does not mean that the materials remain in the area; apparently they are transported near the bottom. The models used by Schubel et al. and by Terwindt are both of the superposition type.

XIII. PHYSICAL MODELS

Hydraulic scale models can be very useful in water pollution problems. General recognition of this fact is relatively recent; the majority of the relevant literature is from the seventies,[17,164-176] but there are some earlier references as well.[177-179] Somewhat less recent are the first applications of hydraulic models in problems of salt intrusion,[180] but one might remark that it is with difficulty that sea water can be seen as a pollutant in a book on pollutants in the sea. It can further be noted that all applications are in physical models of bays, harbors, and estuaries, although some applications indeed concern an area of the open sea,[168,181,182] although then it regards a region that is part of a model of a bay or estuary, mainly serving to obtain proper boundary conditions for the actual range of interest.

The observed facts are connected with the circumstance that except for specific initial mixing problems, the physical models used for pollution studies so far were never constructed with the main or sole purpose of studying pollution and/or dispersion. This is likely so for economic reasons. One of the disadvantages compared to mathematical modeling is that in the latter case a choice between very simple, intermediate, and most detailed modeling is available while the construction of an adequate physical model is always a costly and time-consuming affair, certainly so when the model must compete with mathematical modeling. However, as soon as the physical model is available thanks to other applications, the competition is very well possible because of three-dimensionality, great detail, absence or numerical problems and numerical inaccuracies.

For large open-sea areas no physical model studies of dispersion are known for the simple reason that physical models for such regions hardly exist. The reasons that physical models on such a scale for other applications have been built rather seldomly[183-187] would just as well apply for the purpose of dispersion studies. A minimum requirement would be that the main currents and general circulations be adequately represented, requiring a rotating model to account for Coriolis forces, while at the same time the effect of the wind stress at the surface[188] can hardly be omitted in such a model. The latter effect is more important in large open-sea areas than in bays and estuaries (where the most important wind effects can usually be incorporated in the boundary conditions). Of course, for the same reason, local wind stress must also be accounted for in mathematical models of large sea areas.[4,8,120-123,138]

The above does not mean that in cases where physical models are available they are adequate for all dispersion problems within the concerned region. As we have seen in the foregoing sections, dispersion is a phenomenon governed by velocity patterns at various scales, induced by various mechanisms and boundary conditions, including bottom topography. For each problem it has to be considered which scales and mechanisms are dominant for the (final) pollutant distribution(s) of interest.[175,189] This in turn determines the requirements for a possible hydraulic model for reproduction of the dispersion phenomenon. Generally speaking, the larger the scales of interest, the less important the smaller-scale characteristics of the flow. Sometimes, two different mechanisms may partly act at the same scale range. As an example, shear dispersion and flow patterns induced by bottom topography can be mentioned. In cases where the bottom topography is rather pronounced, as in many estuaries, shear diffusion may become relatively unimportant on a scale where it is still dominant in other areas such as shallow regions of the open sea with a relatively flat bottom. Shear dispersion associated with vertical shear largely depends upon the small-scale vertical mixing and the latter may therefore notably affect horizontal mixing on much larger scales, especially when no pronounced topography-induced horizontal current patterns do occur. For reasons of this kind, one cannot tell by model scales and distortion factors alone what result is to be expected in simulations of dispersion phenomena. To find out which phenomena and scales are dominating, empirical investigations both in model and prototype can be useful.[17] If it is well established which patterns or mechanisms are dominant, one can usually determine what can be expected by using the known model laws for the pertinent flow phenomena.[189,190] For example, if density gradient flows are important, the densimetric Froude number as a model criterion for these flows becomes a criterion for the reproduction of the associated dispersion as well. So, in fact, there are no separate scaling laws for dispersion; the laws to be obeyed are those which regard the reproduction of the flow phenomena dominant in the concerned dispersion problem.[169,175,176]

It is not always necessary to reproduce the dispersion process on the same time scale as the flow phenomena for which the model was designed. Fischer[190] mentions the fact that in salt intrusion studies, an alteration of the (properly scaled) fresh-water discharge may result in a correct salinity distribution.

In more or less stratified estuaries it may be necessary to enhance vertical mixing in the model for obtaining proper salinity distributions. This can often be done successfully by changing the shape of the artificial roughness (vertical rods instead of bottom roughness) in such a way that the resistance for the mean flow is not affected.

A good reproduction of salinity in a bay or estuary does not guarantee a correct image of the small-scale part of a pollutant distribution from a small source, since the small-scale horizontal mixing is not important for the salinity, which intrudes over the entire width. Fischer[190] remarks that a physical model can be used to obtain useful comparisons, e.g., for the choice of outfall sites, without giving correct absolute values of concentrations. This may also apply to mathematical models (compare the remarks on sensitivity analysis in Section XI).

It can be finally mentioned that hydraulic models are often used as an aid in the design of diffusers for sewage outfalls and cooling water in marine areas,[191,192] in relation to problems of initial mixing and in the case of cooling water, "recirculation". In these "near-field models" the effects induced by the waste fluid itself (density differences, induced currents, jet diffusion) are the primary topics. The boundary conditions for the hydraulic near-field models are often obtained from a second model, in this connection called "far-field model",[58,176,192,193] which is usually a mathematical model. For the complicated three-dimensional near-field studies, including density effects, shoreline attachment, etc., hydraulic models are, at the present state of technology, in fact the only appropriate model type.

ACKNOWLEDGMENTS

I wish to thank Dr. G. Abraham and Dr. R. A. Pasmanter for their valuable comments and references and Maaike van Huisstede for typing and retyping with everlasting care and patience.

NOTATIONS

A_w	Surface heat loss coefficient	energy $\times$ area^{-1} $\times$ time^{-1}
$A(x)$	(Wet) cross section of estuary	area
a	Coefficient in the expression $ar^m f(t)$ for a certain class of generalized "diffusion coefficients"	length^{2-m} $\times$ time^{-1}
α	Exponent expressing increasing length of random displacements	—
α	Only in citation of Okubo (Section V): stretching deformation	time^{-1}
α_r	Reaeration coefficient	time^{-1}
β	Exponent in expressions of the form: max. conc. $\propto t^{-\beta}$	—
c, C	Concentration	quantity $\times$ volume^{-1} *
C	Only in citation of Schönfeld (Section IV): constant factor in concentration formula	depends on definition of s
$\bar{c}$	Time-average of concentration; ensemble average of concentration	quantity $\times$ volume^{-1}
c'	Concentration fluctuation	quantity $\times$ volume^{-1}
$c_2, \bar{c}_2, C_2$	Two-dimensional (vertically integrated) concentrations	quantity $\times$ area^{-1}
c_p	Concentration in p-dimensional space	quantity $\times$ length^{-p}
c_o	Coefficient, used when maximum concentration is equal to $Mc_o t^{-\beta}$	length^{-p} $\times$ time$^\gamma$
c_w	Specific heat	energy $\times$ mass^{-1} $\times$ degree^{-1}
γ	(Positive) exponent if one can write $K \propto L^\gamma$	—
$\delta(\ldots)$	Dirac "function"	—
e	2.71828 . . .	—
ε	Energy dissipation per unit mass	energy $\times$ mass^{-1} $\times$ time^{-1}
ε_j	Anisotropy factor	—
f( )	Function of	—
F(t)	Specific decay function ($\neq e^{-t/\tau}$)	—
ϕ	Phase (of eddy movement)	(radians)
h, H	Layer thickness or extension of vertical line source	length
h(t)	Time-dependent thickness of contaminated layer	length
h	Only in citation of Okubo (Section V): shearing deformation	time^{-1}
i, j	Integers for numbering	—
K	One-dimensional or isotropic (turbulent) diffusion coefficient, exchange coefficient, dispersion coefficient	length^2 $\times$ time^{-1}
k	Rate of decay ($1/\tau$)	time^{-1}
$\overset{\rightrightarrows}{K}$	(Turbulent) diffusion tensor	length^2 $\times$ time (components)

---

* Quantity may be mass but also amount of heat, number of bacteria, etc.

$K_{xx} K_{xy}$ $K_{xz} K_{yx}$ $K_{yy} K_{yz}$ $K_{zx} K_{zy}$ K_{zz}	Components of $\overset{\Rightarrow}{K}$ in three dimensions	length² × time⁻¹
K_x, K_y, K_z	Components of $\overset{\Rightarrow}{K}$ for principal axes X, Y, Z	length² × time⁻¹
K_x, K_y	(Also) turbulent diffusion coefficients in two dimensions, horizontal	length² × time⁻¹
K_z, K_v	(Also) turbulent diffusion- or exchange-coefficient in vertical direction	length² × time⁻¹
K_h	(Isotropic) horizontal exchange coefficient	length² × time⁻¹
$K(x)$	Longitudinal exchange coefficient in estuary (tide averaged)	length² × time⁻¹
K_g	(Generalized) diffusion coefficient for patch diffusion	length² × time⁻¹
$\varkappa$	Coefficient of molecular diffusion	length² × time⁻¹
l	Linear measure of patch size, like $(n \times)\sigma_{rc}$	length
L	Linear measure of scale of turbulence or "diffusion"	length
λ_j	"Eddy scales"	length
$M, M_o, M(0)$	Injected quantity of pollutant	quantity
$M(t)$	Decaying quantity of pollutant	quantity
M_D	Total oxygen deficit	mass
m	Exponent in the expression $ar^m f(t)$ for a specific class of generalized "diffusion coefficients"	—
n	Non-negative integer	—
P	Diffusion velocity according to Joseph and Sendner	length × time⁻¹
p	Number of spatial dimensions	—
π	3.14159 . . .	—
$\prod$	Product of factors, numbered i = 1, 2, . . .	—
Q	Released quantity (in Okubo's expression for c_{max} in the case of uniform shear)	quantity
$Q(t)$	Release function of contaminant	quantity × time⁻¹
r	Absolute value of radius vector in polar coordinate system (1, 2 and 3 dimensions)	length
ϱ	Density	mass × volume⁻¹
ϱ	Only in citation of Okubo (Section V): ratio of minor axis to major axis of patch	—
s	Only in citation of Schönfeld (Section IV): concentration	quantity × volume⁻¹ or quantity × area⁻¹
$\sum_i$	Sum of terms, numbered i = 1, 2, . . .	—
σ	Standard deviation; square root of (generalized) variance σ²	length
σ_{rc}^2	Generalized variance as defined by Okubo	length²
T	Time interval for averaging of velocities or concentrations	time
t	Time; in case of instantaneous release, time after injection	time

t_s	Time shift	time
θ	Age of individual patch in superposition model	time
τ	Time in $U_i(\tau)$ ("Lagrangian velocity")	time
τ	Time constant of decay	time
U	Systematic (T-average) velocity in x-direction	length × time⁻¹
u	(Total) velocity in x-direction	length × time⁻¹
$U_i(\tau)$	Velocity history of patch center ("Lagrangian velocity") in i-direction (i = 1 ↔ x, etc.)	length × time⁻¹
u'	Fluctuation u-U	length × time⁻¹
V	Systematic (T-average) velocity in y-direction	length × time⁻¹
$V_i(t)$	Derivative of center line of plume	length × time⁻¹
v	(Total) velocity in y-direction	length × time⁻¹
v'	Fluctuation v-V	length × time⁻¹
W	Systematic (T-average) velocity in z-direction	length × time⁻¹
w	(Total) velocity in z-direction	length × time⁻¹
w'	Fluctuation w-W	length × time⁻¹
W	Only in citation of Schönfeld (Section IV): Schönfeld's diffusion velocity	length × time⁻¹
w_p	Falling rate of particle in still water	length × time⁻¹
x	Spatial coordinate (horizontal)	length
x_i	Spatial coordinate	length
x_{M_i}	Moving center of patch (short notation)	length
(x_M, y_M)	Moving center of patch in two dimensions (horizontal)	length
$\xi_1, \xi_2, \ldots$	Spatial coordinates, e.g., in a moving system	length
ξ	(= ξ_1) Spatial coordinate	
y	Spatial coordinate (horizontal)	length
η	(= ξ_2) Horizontal coordinate, e.g., in a moving system	length
η	Only in citation of Okubo (Section V): vorticity	time⁻¹
z	Vertical coordinate	length
ζ	(= ζ_3) Vertical coordinate, e.g., in a moving system	length

REFERENCES

1. **Abraham, G., van Os, A. G., and Verboom, G. K.,** Mathematical modeling of flows and transports of conservative substances; requirements for predictive ability, in Symp. Predictive Abilities of Surface Water Flow and Transport Models, Berkeley, Calif., 1980.
2. **Kullenberg, G.,** Physical processes, in *Pollutant Transport and Transfer in the Sea,* Vol. I, CRC Press, Boca Raton, Fla., 1982, Chap. 1.
3. **Leendertse, J. J.,** Aspects of a Computation Model for Long-Period Water-Wave Propagation, Rept. No. RM-5294-PR, The Rand Corporation, Santa Monica, Calif., 1967.
4. **Leendertse, J. J.,** A Water-Quality Simulation Model for Well Mixed Estuaries and Coastal Seas, Vol. 1, Rept. no. RM-6230-RC, The Rand Corporation, Santa Monica, Calif., 1970.
5. **Leendertse, J. J. and Gritton, E. C.,** A Water-Quality Simulation Model for Well Mixed Estuaries and Coastal Seas, Vol. II, Rept. no. R-708-NYC, The New York City Rand Institute, New York, 1971.
6. **Leendertse, J. J. and Gritton, E. C.,** A Water-Quality Simulation Model for Well Mixed Estuaries and Coastal Seas, Vol. III to V, Rept. no. R-709-NYC, R-1009-NYC, R-1010-NYC, The New York City Rand Institute, New York, 1971, 1972.
7. **Johnson, C. N. and Leendertse, J. J.,** Input Data Processor (Version 3) of the Two-Dimensional Water-Quality Simulation Model, Working Note WN-9233-USGS/NETH., The Rand Corporation, Santa Monica, Calif., 1975 (restricted).
8. **Maier-Reimer, E.,** Hydrodynamisch-Numerische Untersuchungen zu Horizontalen Ausbreitungs- und Transportvorgangen in der Nordsee, Dissertation, Mitt. Inst. für Meereskunde der Universität Hamburg, XXI, 1973.
9. **Hinze, J. O.,** *Turbulence,* McGraw Hill, New York, 1959.
10. **Batchelor, G. K.,** *The Theory of Homogeneous Turbulence,* Cambridge University Press, New York, 1970.
11. **Corrsin, S.,** Turbulent flow, *Am. Sci.,* 49, 300, 1961.
12. **Tennekes, H. and Lumley, J. L.,** *A First Course in Turbulence,* MIT Press, Cambridge, Mass., 1972.
13. **Csanady, G. T.,** *Turbulent Diffusion in the Environment,* D. Reidel, Boston, 1973.
14. **Okubo, A.,** Horizontal and vertical mixing in the sea, in *Impingement of Man on the Oceans,* Hood, D. W., Ed., Interscience, New York, 1971, 89.
15. **Zimmerman, J. T. F.,** Topographic generation of residual circulation by oscillatory (tidal) currents, *Geophys. Astrophys. Fluid Dyn.,* 11, 35, 1978.
16. **Riepma, H. W.,** Spatial Variability of Residual Currents in an Area of the Southern North Sea, ICES, Hydrogr. Comm., CM 1977/C:43 (restricted), Hydrography Committee, ICES, 1977.
17. Delft Hydraulics Laboratory. Dispersion in Estuaries. Comparison of In-Situ and Small Scale Model Measurements of the Brouwershavense Gat - Grevelingen Estuary (in Dutch, with English abstract), M 1010, Delft Hydraulics Laboratory, Delft, Netherlands, 1974.
18. **Wolff, P. M., Hansen, W., and Joseph, J.,** Investigation and prediction of dispersion of pollutants in the sea with hydrodynamical numerical (HN) models, in *Marine Pollution and Sea Life,* Ruivo, M., Ed., Fishing News (Books) Ltd., London, 1972, 146.
19. North Sea Model, 1977/JAB, Danish Hydraulic Institute, Copenhagen, 1977.
20. **Eggink, H. J.,** The Estuary as a Recipient of Large Quantities of Waste Waters, (in Dutch, with English summary), thesis, Wageningen, 1965.
21. **Uncles, R. J. and Radford, P. J.,** Seasonal and spring-neap tidal dependence of axial dispersion coefficients in the Severn estuary, *J. Fluid Mech.,* 98(4), 703, 1980.
22. **Dorrestein, R.,** A method of computing the spreading of matter in the water of an estuary, in Proc. Symp. Disposal of Radioactive Wastes, International Atomic Energy Agency, Vienna, 1960, 163.
23. **Dorrestein, R. and Otto, L.,** On the mixing and flushing of the water in the Ems-estuary, *Verh. K. Ned. Geol. Mijnbouwkd. Genoot.,* 19, 83, 1960.
24. **van Dam, G. C. and Schönfeld, J. C.,** Experimental and Theoretical Work in the Field of Turbulent Diffusion Performed with Regard to the Netherlands' Estuaries and Coastal Regions of the North Sea, Physics Division, Rijkswaterstaat, The Netherlands, MFA 6807/General Assembly I.U.G.G., Berne, 1967.
25. **van Dam, G. C., Suijlen, J. M., and Brunsveld van Hulten, H. W.,** A one-dimensional water quality model based upon finite difference methods, in Proc. Symp. Systems and Models in Air and Water Pollution, London, 1976, paper 12.
26. **Chatwin, P. C. and Sullivan, P. J.,** The relative diffusion of a cloud of passive contaminant in incompressible turbulent flow, *J. Fluid Mech.,* 91, 337, 1979.
27. **van Dam, G. C.,** Preliminary Results of First Measurements and Computations of Diffusive Spread in the North Sea at Short Distances From the Coast of Holland, Rept. no. FA-1963-1, Physics Division, Rijkswaterstaat, The Netherlands, 1963 (in Dutch).

28. **Joseph, J., Sendner, H., and Weidemann, H.,** Untersuchungen über die horizontale Diffusion in der Nordsee, *Dtsch. Hydrogr. Z.,* 2, 57, 1964.
29. **van Dam, G. C.,** Horizontal diffusion in the North Sea near The Netherlands' coast in connection with waste disposal, in 11th Int. Congress, IAHR, Leningrad, 1965.
30. **Barrett, M. J., Munro, D., and Agg, A. R.,** Radiotracer dispersion studies in the vicinity of a sea outfall, in *Proc. 4th Int. Conf. on Water Pollution,* Pergamon Press, New York, 1969, 863.
31. **Talbot, J. W. and Talbot G. H.,** Diffusion in shallow seas and in English coastal and estuarine waters, *Rapp. P. V. Réun. Cons. Int. Explor. Mer,* 167, 93, 1974.
32. **Weidemann, H.,** *The ICES Diffusion Experiment Rheno 1965,* ICES, Denmark, 1973.
33. **Talbot, J. W.,** Interpretation of diffusion data, in *Proc. Symp. Discharge of Sewage from Sea Outfalls,* London, 1974, 321.
34. **Okubo, A.,** Some speculations on oceanic diffusion diagrams, *Rapp. P. V. Reun. Cons. Int. Explor. Mer,* 167, 77, 1974.
35. **Joseph, J. and Sendner, H.,** Über die horizontale Diffusion im Meere, *Dtsch. Hydrogr. Z.,* 11, 49, 1958.
36. **Okubo, A.,** A New Set of Oceanic Diffusion Diagrams, Tech. Rept. 38, Ref. 68-6, Chesapeake Bay Institute, The Johns Hopkins University, Baltimore, 1968.
37. **Okubo, A.,** A review of theoretical models for turbulent diffusion in the sea, *J. Oceanogr. Soc. Jpn.,* 20, 286, 1962.
38. **van Dam, G. C.,** Some Basic Concepts in the Theory of Turbulent Diffusion. An Elementary Approach, Rept. no. MFA 6901^E, Physics Division, Rijkswaterstaat, The Netherlands, 1969.
39. **Kullenberg, G.,** Investigations on dispersion in stratified vertical shear flow, *Rapp. P. V. Reun. Cons. Int. Explor. Mer,* 167, 86, 1974.
40. **van Dam, G. C. and Sydow, J. S.,** A Diffusion Experiment Near the Dutch Coast 10 km. off Ter Heyde, Rept. no. MFA 7003, Physics Division, Rijkswaterstaat, The Netherlands, 1970 (in Dutch).
41. Diffusion Mechanisms in Pipes, Canals, Rivers, Estuaries and Coastal Seas (Flow Without Density Effects) Rept. no. R 895-2, Delft Hydraulics Laboratory, 1978, (in Dutch).
42. **van Dam, G. C.,** Sensitivity Analysis with Regard to Various Transport Mechanisms, Texel 1978, J. T. F., Zimmerman, Ed., Netherlands Institute for Sea Research, 1978, 69 (in Dutch).
43. **van Dam, G. C. and Davids, J. A. G.,** Radioactive waste disposal and investigations on turbulent diffusion in the Netherlands' coastal areas, Proc. Symp. Disposal of Radioactive Wastes into Seas, Oceans and Surface Waters, Vienna, 1966, 233.
44. **Okubo, A.,** Oceanic diffusion diagrams, *Deep Sea Res.,* 18, 789, 1971.
45. **Kolmogorov, A. N.,** The local structure of turbulence in an incompressible viscous fluid for very large Reynolds' numbers, *C. R. Dokl. Acad. Sci. URSS,* 30, 301, 1941.
46. **Monin, A. S. and Yaglom, A. M.,** *Statistical Fluid Mechanics,* MIT Press, Cambridge, Mass., 1971.
47. **Tennekes, H.,** The exponential Lagrangian correlation function and turbulent diffusion in the inertial subrange, *Atmos. Environ.,* 13, 1565, 1979.
48. **Pasmanter, R. A.,** Diffusion in highly turbulent fluids, *Phys. Lett.,* A75, 366, 1980.
49. **Nihoul, J. C. J.,** The turbulent ocean, in *Marine Turbulence, Proc. 11th Int. Liège Colloq. Ocean Hydrodynamics,* Nihoul, J. C. J., Ed., Elsevier, Amsterdam, 1980, 1.
50. **Okubo, A. and Ozmidov, R. V.,** Empirical dependence of the coefficient of horizontal turbulent diffusion in the ocean on the scale of the phenomenon in question, *Izv. Atmospheric Oceanic Phys. Ser.,* 6, 534, 1970.
51. **Ozmidov, R. V.,** Energy distribution between oceanic motions of difference scales, *Izv. Atmospheric Oceanic Phys. Ser.,* 1, 439, 1965.
52. **Bowden, K. F.,** Turbulence. II, *Oceanogr. Mar. Biol. Annu. Rev.,* 8, 11, 1970.
53. **Neumann, G. and Pierson, W. J.,** *Principles of Physical Oceanography,* Prentice-Hall, Englewood Cliffs, N.J., 1966.
54. **Nihoul, J. C. J.,** Passive dispersion models, in *Modeling of Marine Systems,* Nihoul, J. C. J., Ed., Elsevier, Amsterdam, 1975, chap. 3.
55. **Ozmidov, R. V.,** On the calculation of horizontal turbulent diffusion of the pollutant patches in the sea, *Dokl. Akad. Nauk, SSSR,* 120, 761, 1958.
56. **Ozmidov, R. V.,** *Horizontal Turbulence and Turbulent Exchange in the Ocean,* Publ. House "Science", Moscow, 1968 (in Russian).
57. **Okubo, A. and Pritchard, D. W.,** unpublished note, see Pritchard, D. W., The application of existing oceanographic knowledge to the problem of radioactive waste disposal into the sea, in Proc. Symp. Disposal of Radioactive Wastes, International Atomic Energy Agency, Vienna, 1960, 229.
58. **Carter, H. H.,** Prediction of far-field exclusion areas and effects, *Proc. Symp. Discharge of Sewage from Sea Outfalls,* London, 1974, 363.
59. **Suijlen, J. M.,** Turbulent Diffusion in the IJsselmeer near Medemblik, Measured by Means of Rhodamine B, Rept. no. FA 7501, Physics Division, Rijkswaterstaat, The Netherlands, 1975 (in Dutch).

60. **Murthy, C. R.,** Horizontal diffusion characteristics in Lake Ontario, *J. Phys. Oceanogr.,* 6, 76, 1976.
61. **Kirwan, A. D., Jr., McNally, G. J., Reyna, E., and Merrell, W. J., Jr.,** The near-surface circulation of the eastern North Pacific, *J. Phys. Oceanogr.,* 8, 937, 1978.
62. **Schönfeld, J. C.,** Integral diffusivity, *J. Geophys. Res.,* 67, 8, 1962.
63. **van Dam, G. C.,** A two-dimensional water quality model based upon the superposition principle, Proc. Symp. Systems and Models in Air and Water Pollution, London, 1976, paper 11.
64. **Monin, A. S.,** An equation of turbulent diffusion, *Dokl. Akad. Nauk, SSSR,* 105, 256, 1955 (in Russian).
65. **Berkowicz, R. and Prahm, L. P.,** On the spectral turbulent diffusivity theory for homogeneous turbulence, *J. Fluid Mech.,* 100(2), 433, 1980.
66. **Berkowicz, R. and Prahm, L. P.,** Generalization of K-theory for turbulent diffusion. I. Spectral turbulent diffusivity concept, *J. Appl. Meteorol.,* 18, 226, 1979.
67. **Kullenberg, G.,** Experimental techniques, in *Pollutant Transport and Transfer in the Sea,* Vol. I, CRC Press, Boca Raton, Fla., 1982, chap. 3.
68. **van Dam, G. C.,** Dispersion of Dissolved or Suspended Matter Released Into the North Sea Near Wijk aan Zee, 3 km. From the Shore, Rept. no. MFA 6812, Physics Division, Rijkswaterstaat, The Netherlands, 1968 (in Dutch with English summary).
69. **van Dam, G. C.,** Interpretation of Visual Area of Dye Patches on Aerial Photographs Rept. no. FA 7701, Physics Division, Rijkswaterstaat, The Netherlands, 1977 (in Dutch).
70. **Okubo, A.,** Horizontal Diffusion From an Instantaneous Point-Source Due to Oceanic Turbulence, Tech. Rept. 32, Ref. 62-22, Chesapeake Bay Institute, The Johns Hopkins University, Baltimore, 1962.
71. **Suijlen, J. M.,** Working Notes 78-FA-175 and 78-FA-178, Physics Division, Rijkswaterstaat, The Netherlands (to be published shortly), 1978.
72. **Taylor, G. I.,** Dispersion of soluble matter in solvent flowing slowly through a tube, *Proc. R. Soc. London Ser. A,* 219, 186, 1953.
73. **Taylor, G. I.,** The dispersion of matter in turbulent flow through a pipe, Proc. R. Soc. London Ser. A, 223, 446, 1954.
74. **Huang, C. H.,** A theory of dispersion in turbulent shear flow, *Atmos. Environ.,* 13, 453, 1979.
75. **Okubo, A.,** A note on horizontal diffusion from an instantaneous source in a nonuniform flow, *J. Oceanogr. Soc. Jpn.,* 22, 35, 1966.
76. **Okubo, A.,** The effect of shear in an oscillatory current on horizontal diffusion from an instantaneous source, *Int. J. Oceanol. Limnol.,* 1, 194, 1967.
77. **Novikov, E. A.,** Concerning a turbulent diffusion in a stream with a transverse gradient of velocity, *J. Appl. Math. Mech.,* 22, 576, 1958.
78. **Bowden, K. F.,** Horizontal mixing in the sea due to a shearing current, *J. Fluid Mech.,* 21, 83, 1965.
79. **Schönfeld, J. C.,** The mechanism of longitudinal diffusion in a tidal river, Proc. Symp. Tidal Rivers, Helsinki, 1960.
80. **Okubo, A. and Carter, H. H.,** An extremely simplified model of the 'shear effect' on horizontal mixing in a bounded sea, *J. Geophys. Res.,* 71, 5267, 1966.
81. **Thacker, W. C.,** A solvable model of 'shear dispersion', *J. Phys. Oceanogr.,* 6, 66, 1976.
82. **Carter, H. H. and Okubo, A.** A Study of the Physical Process of Movement and Dispersion in the Cape Kennedy Area, Final Rept. under U.S. AEC Contract AT(30-1)-2973 Ref. 65-2, Chesapeake Bay Institute, The Johns Hopkins University, Baltimore, 1965.
83. **Okubo, A.,** Some remarks on the importance of the "shear effect" on horizontal diffusion, *J. Oceanogr. Soc. Jpn.,* 24, 60, 1968.
84. **Harremoës, P.,** Prediction of pollution from planned sewage outlets, *Tek. Kemi. Aikak.,* 21, 6, 251, 1964.
85. **Harremoës, P.,** Prediction of pollution from planned wastewater outfalls, *J. Water Pollut. Control Fed.,* 38, 1323, 1966.
86. **Frenkiel, F. N.,** Turbulent diffusion, *Adv. Appl. Mech.,* 3, 61, 1953.
87. **Gifford, F., Jr.,** Statistical properties of a fluctuating plume dispersion model, *Adv. Geophys.,* 6, 117, 1959.
88. **Schönfeld, J. C.,** Turbulent Diffusion (Continuous Release from a Source in the Sea), Rept. no. MFA6411, Physics Division, Rijkswaterstaat, The Netherlands, 1964 (in Dutch).
89. **Carter, H. H. and Okubo, A.,** Comments on paper by G. T. Csanady, "Accelerated diffusion in the skewed shear flow of lake currents," *J. Geophys. Res.,* 71, 5012, 1966.
90. **Okubo, A. and Karweit, J. J.,** Diffusion from a continuous source in a uniform shear flow, *Limnol. Oceanogr.,* 14, 514, 1969.
91. **Nuclear Energy Agency,** A Hazard Assessment for Radioactive Waste Disposal into the North East Atlantic, Report by an NEA Group of Experts, Paris, 1973.

92. **van Dam, G. C.**, Horizontal Two-Dimensional Mathematical Models of Dispersion in Turbulent Waters, Especially by the So-Called Superposition Method, Rept. no. FA7302, Physics Division, Rijkswaterstaat, The Netherlands, 1973 (in Dutch).

93. **Wen-Hsiung, L. W.**, *Differential Equations of Hydraulic Transients, Dispersion and Groundwater Flow. Mathematical Methods in Water Resources*, Prentice-Hall, Englewood Cliffs, New Jersey, 1972.

94. **Fischer, H. B., List, E. J., Koh, R. C. Y., Imberger, J. and Brooks, N. H.**, *Mixing in Inland and Coastal Waters*, Academic Press, New York, 1979.

95. **Pasquill, F.**, *Atmospheric Diffusion. A Study of the Dispersion of Windborne Material from Industrial and Other Sources*, 2nd ed., Ellis Harwood Ltd., Chichester 1974.

96. **Crank, J.**, *The Mathematics of Diffusion*, Clarendon Press, Oxford, 1956.

97. **Carslaw, H. S. and Jaeger, J. C.**, *Conduction of Heat in Solids*, 2nd ed., Clarendon Press, Oxford, 1959.

98. **van Dam, G. C.**, unpublished note 15 77-FA, Physics Division, Rijkswaterstaat, The Netherlands, July 1977 (in Dutch).

99. **Volker, W. F.**, personal communication, Data Processing Division, Rijkswaterstaat, 1973.

100. Commissie afvalvraagstuk zuidelijk deel randstad Holland, First report, 1965 (in Dutch).

101. Delft Hydraulics Laboratory, Gold Coast Queensland Ocean Outfall Sewers. Bacteriological and Diffusion Aspects, Rept. R398-1, Delft Hydraulics Laboratory, 1970.

102. **van Dam, G. C.**, The Hague outfall, in Proc. Symp. Discharge of Sewage from Sea Outfalls, London, 1974, 393.

103. **Brooks, N. H.**, Diffusion of sewage effluent in an ocean-current, in *Proc. 1st Int. Conf. Waste Disposal in the Marine Environment, 1959*, Pearson, E. A., Ed., Pergamon Press, Elmsford, N.Y., 1960, 246.

104. **Siemons, J.**, Numerical methods for the solution of diffusion advection equations, Publ. 88 Delft Hydraulics Laboratory, 1970.

105. **Weare, T. J.**, Finite element or finite difference methods for the two-dimensional shallow water equations?,. *Comp. Meth. Appl. Mech. Eng.*, 7, 351, 1976.

106. **Cheng, R. T.**, Modeling of hydraulic systems by finite-element methods, *Adv. Hydrosci.*, 11, 207, 1978.

107. **Ramming, H. G.**, Numerical investigations of the influence of coastal structures upon the dynamic off-shore process by application of a nested tidal model, in *Hydrodynamics of Estuaries and Fjords, Proc. 9th Int. Liège Colloq. Ocean Hydrodynamics*, Nihoul, J. C. J., Ed., Elsevier, Amsterdam, 1978, 315.

108. **Gärtner, S.**, Zur Berechnungen von Flachwasserwellen und instationären Transportprozessen mit der Methode der finiten Elemente, *Fortschr. Ber. VDI, Z.*, 4, 30, 1977.

109. **Lam, D. C. L.**, Comparison of finite-element and finite-difference methods for nearshore advection-diffusion transport models, in *Finite Elements in Water Resources*, Gray, W. G., Pinder, G. F., and Brebbia, C. A., Eds., Pentech Press, London, 1977, 115.

110. **Matsuda, Y.** A water pollution prediction system by the finite element method, *Adv. Water Res.*, 2, 27, 1979.

111. **Brebbia, C. A. and Partridge, P. W.**, Finite element simulation of water circulation in the North Sea, *Appl. Math. Modeling* 1, 101, 1976.

112. **Wang, J. D.**, Real-time flow in unstratified shallow water, *J. Waterway Port Coastal Ocean Div., Proc. ASCE*, 104, 53, 1978.

113. **Thacker, W. L.**, Comparison of finite-element and finite-difference schemes, *J. Phys. Oceanogr.*, 8, 676, 1978.

114. **Norrie, D. H., and de Vries, G.**, *The Finite Element Method*, Academic Press, New York, 1973.

115. **Vreugdenhil, C. B. and Voogt, J.**, Hydrodynamic Transport Phenomena in Estuaries and Coastal Waters. Scope of Mathematical Models, Publ. 155, Delft Hydraulics Laboratory, 1975.

116. **Lam, D. C. L. and Simons, T. J.**, Numerical computations of advective and diffusive transports of chloride in Lake Erie, 1970, *J. Fish. Res. Board Canada*, 33, 537, 1976.

117. **Okubo, A.**, Remarks on the use of "diffusion diagrams" in modeling scale-dependent diffusion, *Deep Sea Res.*, 23, 1213, 1976.

118. **Leendertse, J. J. and Liu, S-K.**, A Three-Dimensional Model for Estuaries and Coastal Seas, Vol. IV, Rept. no. R2187-OWRT, The Rand Corporation, Santa Monica, Calif., 1977.

119. **Rodi, W.**, Turbulence models and Their Application in Hydraulics — A State of the Art Review, Rep. SFB 80/T/127, University Karlsruhe, Karlsruhe, Federal Republic of Germany, 1978.

120. **Hansen, W.**, The computation of tides and storm surges and its application to problems of coastal engineering, Paper 3.4 presented at 9th Conf. Coastal Engineering, Lisbon, 1964.

121. **Brettschneider, G.**, Modelluntersuchungen der Gezeiten der Nordsee unter Anwendung des hydro-dynamisch-numerischen Verfahrens, dissertation, Mitteilungen des Instituts für Meereskunde der Universität Hamburg VIII, Hamburg, Federal Republic of Germany, 1967.

122. **Maier-Reimer, E.,** Zum Einfluss eines mittleren Windschubes auf die Restströme der Nordsee, *Dtsch. Hydrog. Z.,* 6, 253, 1975.

123. **Maier-Reimer, E.,** Residual circulation in the North Sea due to the M_2-tide and mean annual wind stress, *Dtsch. Hydrogr. Z.,* 30, 69, 1977.

124. **Ivanov, K., and Filippov. Y. G.,** Propagation of a dynamically passive impurity in the surface layer of the sea, *Oceanology,* 18, 275, 1978.

125. **Thompson, R.** Numeric calculation of turbulent diffusion, *Q. J. R. Meteorol. Soc.,* 97, 93, 1971.

126. **Bugliarello, G. and Jackson, E. D.,** Random walk study of convective diffusion, *J. Eng. Mech. Div. Proc. ASCE,* 90, 49, 1964.

127. **Einstein, A.,** Über die von der molekularkinetischen Theorie der Wärme geforderte Bewegung von in ruhenden Flüssigkeiten suspendierten Teilchen, *Ann. Phys. (Leipzig),* 4, 17, 1905.

128. **Chandrasekhar, S.,** Stochastic problems in physics and astronomy, *Rev. Mod. Phys.,* 15, 1, 1943.

129. **Pasmanter, R. A.** Two-Dimensional Random Walk Rept. no. 79-FA-010, Physics Division, Rijkswaterstaat, The Netherlands, 1979.

130. **Kraichnan, R. H.,** Diffusion by a random velocity field, *Phys. Fluids,* 13, 22, 1970.

131. **van Dam, G. C.,** Scale Dependent Dispersion of Distinct Particles in an Artificial Eddy Field, Rept. no. 07 80-FA, Physics Division, Rijkswaterstaat, The Netherlands, 1980.

132. **Bork, I. and Maier-Reimer, E.,** On the spreading of power plant cooling water in a tidal river applied to the river Elbe, *Adv. Water Res.,* 1, 161, 1978.

133. **Delvigne, G. A. L.,** Estmation of Computing Time of a Numerical Particle Model for the North Sea, Rept. no. R1410, Delft Hydraulics Laboratory, 1979 (in Dutch).

134. **den Ouden, T. E. M.,** The Development of a Set of Programs for a Two-Dimensional Particle Model, Rept. no. FA8003, Physics Division, Rijkswaterstaat, The Netherlands, 1980 (in Dutch).

135. Delft Hydraulics Laboratory, Calculation of Flow in a Tidal River Rept. no. R897-I/II/III, Delft Hydraulics Laboratory, 1975—1976 (in Dutch).

136. Delft Hydraulics Laboratory, Computation of Density Currents in Estuaries, Rept. no. R897-IV/V/VI, Delft Hydraulics Laboratory, 1979—1980.

137. Delft Hydraulics Laboratory, Three-Dimensional Circulation Models for Shallow Lakes and Seas Rept. no. R900-I/II/III, Delft Hydraulics Laboratory, 1976, 1979.

138. **Leendertse, J. J., Alexander, R. C., and Liu, S-K.,** A Three-Dimensional Model for Estuaries and Coastal Seas, Vol. I, Rept. no. R-1417-OWRR. The Rand Corporation, Santa Monica, Calif., 1973.

139. **Leendertse, J. J. and Liu, S-K.,** A Three-Dimensional Model for Estuaries and Coastal Seas, Vol. II, Rept. no. R-1764-OWRT, The Rand Corporation, Santa Monica, Calif., 1975.

140. **Liu, S.-K, and Leendertse, J. J.,** A Three-Dimensional Model for Estuarties and Coastal Seas, Vol. VI, Rept. no. R-2405-NOAA, The Rand Corporation, Santa Monica, Calif., 1979.

141. **Riley, G. A.,** Oxygen, phosphate, and nitrate in the Atlantic Ocean, *Bull. Bingham Oceanogr. Collection,* 13, 1, 1951.

142. **Dunster, H. J.,** Introduction to the hazard assessment, in Experimental Disposal of Solid Radioactive Waste into the Atlantic Ocean, ENEA Seminar to Assess Oceanographic and Other Marine Factors, Lisbon, 1967.

143. **Webb, G. A. M. and Grimwood, P. D.,** A Revised Oceanographic Model to Calculate the Limiting Capacity of the Ocean to Accept Radioactive Waste, Rept. no. NRPB-R58, National Radiological Protection Board, Harwell, 1976.

144. **Shepherd, J. G.,** A Simple Model for the Dispersion of Radioactive Wastes Dumped on the Deep-Sea Bed, Rept. 29, Directorate of Fisheries Research, Lowestoft, U.K.,

145. IAEA, The Oceanographic Basis of the IAEA Revised Definition and Recommendations Concerning High-Level Radioactive Waste Unsuitable for Dumping at Sea, International Atomic Energy Agency, Vienna, 1978, 210.

146. **Munk, W. H. and Anderson, E. R.,** Notes on a theory of the thermocline, *J. Mar. Res.,* 7, 276, 1948.

147. Delft Hydraulics Laboratory. Momentum and Mass Transfer in Stratified Flow, Rept. no. R880, Delft Hydraulics Laboratory, 1974.

148. **Abraham, G. and van Dam, G. C.,** On the predictability of waste concentrations, in *Marine Pollution and Sea Life,* Ruivo, M., Ed., Fishing News (Books) Ltd., London, 1972, 135.

149. **Chatwin, P. C. and Sullivan, P. J.,** Measurements of concentration fluctuations in relative turbulent diffusion, *J. Fluid Mech.,* 94, 83, 1979.

150. **Suijlen, J. M.,** A Dye Experiment in the North Sea up to a Very Large Scale, No. C.M. 1980/C:29, ICES, Hydrography Committee, October 1980.

151. Delft Hydraulics Laboratory, Environmental Investigation North Sea Island, Spread of Materials and Heat, Rept. no R1275, Delft Hydraulics Laboratory, 1978.

152. **Timmerman, H.,** On the importance of atmospheric pressure gradients for the generation of external surges in the North Sea, *Dtsch. Hydrog. Z.,* 28, 62, 1975.

153. **Ronday, F. C.**, Modèle Mathématique pour l'Étude des Courants Résiduels dans la Mer du Nord M-R Series 27, Marine Science Directorate, Department of the Environment, Ottawa, Canada, 1972.
154. **Longuet-Higgins, M. S. and Stewart, R. W.**, Radiation stresses in water waves; a physical discussion, with applications, *Deep Sea Res.*, 11, 529, 1964.
155. **Maier-Reimer, E.**, Some effects of the Atlantic circulation and of river discharges on the residual circulation of the North Sea, *Dtsch. Hydrogr. Z.*, 32, 126, 1979.
156. **Otto, L.**, The Mean residual Transport Pattern in the Southern North Sea, No. CM/1970/C:21, ICES, Hydrography Committee (restricted), 1970.
157. **Bolin, B. and Stommel, H.**, On the abyssal circulation of the world ocean. IV. Origin and rate of circulation of deep ocean water as determined with the aid of tracers, *Deep Sea Res.*, 8, 95, 1961.
158. **Otto, L.**, Age Distribution in a Periodically Stratified Basin, No. CM1978/C:43, ICES, Hydrography Committee (restricted), 1978.
159. **Westhoff, J. W., Davids, J. A. G., and van Dam, G. C.**, Dispersion Experiments with Continuous Injection off Petten at 3.5 km from the Shore. I Experiments, Rept. no. MFA 7101, Physics Division, Rijkswaterstaat, The Netherlands, 1971.
160. **Suijlen, J. M.**, On the Dispersion of River Water along the Dutch Coast, Physics Division, Rijkswaterstaat, The Netherlands, report in preparation, 1981.
161. **Groen, P.**, On the residual transport of suspended matter by an alternating tidal current, *Neth. J. Sea Res.*, 3, 564, 1967.
162. **Schubel, J. R., Carter, H. H., Wilson, R. E., Wise, W. M., Heaton, M. G., and Gross, M. G.**, Field Investigations of the Nature, Degree, and Extent of Turbidity Generated by Open-Water Pipeline Disposal Operations, Tech. Rept. D-78-30, U.S. Army Engineer Waterways Experiment Station, Vicksburg, Miss., 1978.
163. Further Investigation of the Sediment-Movement in the Coastal Water Along the Delta Area and the "Schone Kust" in Relation to the Dumping of Dredged Material at "Loswal Noord", Rept. no. W-71.030, Hydraulics Division Deltadienst, Rijkswaterstaat, The Netherlands, 1971 (in Dutch).
164. Strömungs und Durchmischungsmessungen für die Abwassereinleitung in die Ems (Modellversuche), Bundesanstalt für Wasserbau, Hamburg, Federal Republic of Germany, 1970.
165. **Fischer, H. B. and Holley, E. R.**, Analysis of the use of distorted hydraulic models for dispersion studies, *Water Resour. Res.* 7, 46, 1971.
166. **Harleman, D. R. F.**, Physical models, in Estuarine Modeling: an Assessment, Water Pollution Control Research Series, 16070, DZV 02/71 Water Quality Office, U.S. Environmental Protection Agency, Cincinnati, Ohio, 1971, chapt. 5.
167. **Ippen, A. T.**, Review and comments, in Estuarine Modeling: an Assessment, Water Pollution Control Research Series, 16070, DZV 02/71 Water Quality Office, U.S. Environmental Protection Agency, Cincinnati, Ohio, 1971, 489.
168. **Crickmore, M. J.**, Tracer tests of eddy diffusion in field and model, *J. Hydr. Div. Proc. ASCE*, 98, 1737, 1972.
169. **Abraham, G.**, General report (data collection processes and modeling techniques), Vol. 6, Proc. 15th IAHR Congress, Istanbul, 1973, 293.
170. **Higuchi, H. and Sygimoto, T.**, Experimental study of horizontal diffusion due to the tidal current, *Rapp. P. V. Réun. Cons. Int. Explor. Mer*, 167, 177, 1974.
171. **Tamai, N.**, Dispersion models in coastline water with predominant transverse shear, *Coastal Eng. Jpn*, 17, 185, 1974.
172. **Hayakawa, N., Tanabe, H., Takasugi, Y., Takarada, M., Yuasa, I., and Sumimoto, T.**, Hydraulic model study of pollutant dispersion in the Seto inland sea, Vol. 3, Proc. 17th IAHR Congress, Baden-Baden, 1977, 39.
173. **Kato, M., Wada, A., and Tanaka, N.**, Study on reproducibility of diffusion phenomena in hydraulic model for systems of surface and submerged outfall, Vol. 3, Proc. 17th IAHR Congress, Baden-Baden, 1977, 31.
174. **Ohlmeyer, F.**, Scaling of physical models with dispersion problems, Vol. 3, Proc. 17th IAHR Congress, Baden-Baden, 1977, 9.
175. **Di Silvio, G.**, General report (diffusion and dispersion of pollutants under the influence of currents induced by wind, waves, and tides), Vol. 6, Proc. 17th IAHR Congress, Baden-Baden, 1977, 253.
176. **Abraham, G. and Ohlmeyer, F.**, Tidemodelle für Ausbreitungsvorgänge von Wasserinhaltsstoffen und Wärme, in Wasserbauliches Versuchswesen, Mitteilungsheft 4, Deutscher Verband für Wasserwirtschaft (DVWW), 1978, chap. 8.
177. **Simmons, H. B.**, Application and limitations of estuary models in pollution analysis, in *Proc. 1st Int. Conf. Waste Disposal in the Marine Environment*, Pearson, E., Ed., Pergamon Press, New York, 1960.
178. **Harleman, D. R. F.**, Dispersion in hydraulic models, in *Estuary and Coastline Hydrodynamics*, Ippen, A. T., Ed., McGraw Hill, New York, 1966, section 14.4.

179. **Harleman, D. R. F., Holley, E. R., and Huber, W. C.,** Interpretation of water pollution data from tidal estuary models, in Proc. 3rd Int. Conf. Water Pollution, Munich, 1966, paper 3.
180. **U.S. Army Corps of Engineers,** Delaware River Model Study. Hydraulic and Salinity Verification, Rept. no. 1, West. M. 2-337, Waterways Experiment Station, Vicksburg, Miss., 1956.
181. **van Rees, A. J., van der Kuur, P., and Stroband, H. J.,** Experiences with tidal salinity model Europoort, in 13th Int. Conf. Coastal Engineering, Vancouver, 1972.
182. **van Rees, A. J.,** The role of a hydraulic model for operational and basic research, in Waterloopkunde in Dienst van Industrie en Milieu Publ. 110N, Delft Hydraulics Laboratory, 1973, Paper 4 (in Dutch).
183. **Bonnefille, R.,** Note préliminaire sur réalisation d'un modèle réduit tournant du Détroit de Gibraltar, *Cah. Oceanogr.,* 5, 357, 1964.
184. **Bonnefille, R., Braconnot, P., and Donnars, P.,** Projet Gibraltar; Etude sur modèle réduit du Détroit de Gibraltar; rapport no. 1: Le Modèle et son étallonage. Tech. Rep. 33, NATO Subcommittee on Oceanographic Research, Chatou, 1968.
185. **Baker, D. J., Jr. and Robinson, A. R.,** A laboratory model for the general ocean circulation, *Philos. Trans. R. Soc. London Ser. A,* 265, 533, 1969.
186. **Bonnefille, R.,** Contribution théorique et expérimentale à l'éutde du régime des marées, *Bull. Dir. Etudes Rech. Electricite Fr. Ser. A,* 1, 1969.
187. **Ichiye, T.,** Experiments on circulation of the Gulf of Mexico and the Caribbean Sea with a rotating tank, in *Contributions in Oceanography, 1972—1973,* Vol. 16, Contr. no. 538, Texas A & M University, College of Geosciences, College Station, TX., 1973, 271.
188. **Buechi, P. J. and Rumer, R. R.,** Wind induced circulation pattern in a rotating model of Lake Erie, in Proc. 12th Conf. Great Lakes Research, International Association for Great Lakes Research, Ann Arbor, 1969, 406.
189. **Fischer, H. B.,** Mixing and dispersion in estuaries, *Annu. Rev. Fluid Mech.,* 8, 107, 1976.
190. **Fischer, H. B.,** Physical models, in *Mixing in Inland and Coastal Waters,* Academic Press, New York, 1979, chap. 8 (section 8.3).
191. **Koh, R. C. Y.,** Design of ocean wastewater discharge systems, in *Mixing in Inland and Coastal Waters,* Academic Press, New York, 1979, chap. 10.
192. **Ligteringen, H.,** Combined use of hydraulic and mathematical models in the design of a once-through cooling circuit along an estuary, in Waste Heat Management and Utilization Conf., Miami Beach, 1977.
193. **Harleman, D. R. F.,** Fluid mechanics of heat disposal from power generation, *Annu. Rev. Fluid Mech.,* 4, 7, 1972.

Chapter 3

EXPERIMENTAL TECHNIQUES

Gunnar Kullenberg

TABLE OF CONTENTS

I. INTRODUCTION

In recent years a number of powerful techniques have been developed, making it possible to investigate the physical dispersion conditions in coastal areas and the surface layer of the ocean in general. The aim of this chapter is to give an account of some of these techniques and how they may be applied. It is not the intention to discuss chemical or biological observations which it in some cases may be of interest to carry out in conjunction with the physical dispersion studies. For information on these techniques reference is made to relevant chapters in this book as well as to treatises on marine chemistry and biology.

In many applications it is of interest to consider the dispersion of bacteria or virus. Microorganisms are often used as indicators of pollution. Generally, a good knowledge of the environmental conditions (i.e., salinity, temperature, currents and meteorological factors) may make it possible in combination with physical dispersion factors to give a reasonable account of bacterial disappearance in relation for instance to an outfall. However, often it is necessary to carry out field studies in order to obtain a reliable picture of the bacterial disappearance.[1,2]

In other applications the properties of the substance, e.g., oil, will influence the dispersion (see Chapter 7). The possible influence of the dispersing material, e.g., waste water, on the dispersion during an intermediate period following immediately after the initial dispersion may be important in restricted areas with very limited water exchange or in stagnant basins.[3-6] Normally, however, a proper design and siting of the outfall should result in an initial dilution large enough to make the effect insignificant.

The basic observations of the environmental conditions will also be appropriate for calculation of the initial dilution for a given outfall design. Hence, we will concentrate here on the observations required for obtaining an insight into physical dispersion conditions in coastal waters as well as the surface layer of the open ocean. The aim of field investigations is to give a description of the pattern of spreading and transportation in the area and how these features are related to the environmental conditions. From the observations it should be possible to determine mixing parameters and advection velocities and test various models predicting the dispersion and how it relates to the environmental conditions such as currents, stratification, and wind, as well as yield estimates of residence times and maximum expected concentrations reaching specific areas from optional outfall sites. Clearly the design of the field investigations will have to be tailored to the requirements as well as the local conditions. Generally, field investigations should also be combined with various theoretical considerations and possibly numerical modeling.

In general, field studies are expensive and they should be carefully planned so that maximum use can be made of theoretical knowledge and models. The need for field studies will often depend upon the local conditions, such as the form of the coast line and the topography. Open, straight coast lines may imply fairly simple current patterns suitable for mathematical modeling supplemented with few current measurements, whereas a complicated archipelago implying complex circulation patterns may require sophisticated *in situ* tracer studies. In semi-enclosed areas like fjords and estuaries basic information on the dispersion conditions may be obtained by considering the volume, the fresh water runoff, and the topography. Such considerations should always be made prior to launching expensive field experiments.

II. ENVIRONMENTAL OBSERVATIONS

Any field study with the aim of determining the dispersion characteristics in a more or less limited area should include certain basic observations of environmental factors influencing the dispersion. These are primarily the meteorological conditions such as wind, velocity and direction, the temperature and salinity distributions giving the density distribution in the water, and the current distribution. In coastal areas additional observations are often required, for instance of water level and fresh water runoff. The observations should be planned so that a reasonable amount of information on time and space variability is obtained. Both vertical and horizontal mappings of distributions should be included in the program, covering at least seasonal variations. The frequency of observation as well as the spatial resolution is often a matter of available resources. Often it is pertinent to carry out a broad brush survey of the conditions prior to embarking on regular observations in a fixed network of stations. In coastal areas existing records of wind conditions, runoff, and water level variations can be very useful in giving an impression on the range of variability, general types of motion in the area, and the possible influence of runoff on the salinity distribution.

It is important that a reliable coverage is obtained so that also processes occurring during extreme conditions can be detected. Often the effects of extreme conditions cannot be observed directly unless automatic recording instruments are used, and then only if such conditions occur during the period of observation. Hence, existing long records obtained for other purposes can be extremely valuable.

In many areas, in particular estuaries and fjords, the salinity distribution and its time and space variation can give direct information on the mixing conditions (see Chapter 2, Vol. 2). Continuous salinity and temperature profiles are obtained by means of conductivity-temperature-depth (CTD) measuring systems. In some cases it may be more practical to use a small conductivity-temperature sensor lowered directly by hand and measuring at discrete levels. Observations by means of water samples obtained by water bottles have the disadvantage that important features of the vertical distribution are missed, unless a close spacing of the bottles is used. The time needed for such observations is also considerably larger than when *in situ* measurements are used. These techniques are much to be preferred. It should be noted that careful and periodic calibrations of the sensors are necessary for reliable and intercomparable measurements.

Mostly the salinity and temperature observations need to be supplemented with current measurements. The recording automatic instruments which can be moored for considerable periods of time make it possible to obtain continuous and simultaneous observations of current, conductivity (salinity), temperature and pressure (depth) in a network of stations. Such observations give information on water exchange, advection, and mixing conditions. Whenever possible the observations should be combined with water level and wind observations. The combination will make it possible to interpret the current variations and separate effects of winds, tides, and runoff as well as test and calibrate various numerical models. Thus it may be possible to generate operational models of water exchange based on observations which can be obtained in real time and on a continuous basis. An extensive current meter program can clearly not be maintained but for a limited period of time.

The exchange between the coastal zone and the open sea is an important factor which requires field observations covering all relevant dynamical factors for its determination. The combined action of coast line, meteorological forcing, and stratification in the sea is particularly significant in this case.[7]

Recently current meters with no (moving) sensor in the flow have been developed. Such are the electromagnetic current meter, the acoustic current meter measuring the time difference between the passing of a beam across a small length in the opposite direction, and the laser Doppler anometer measuring the Doppler shift of light scattered by particles moving in the flow. Intercomparisons of such meters have shown that they are reliable and yield comparable data.[7a] These instruments are as yet not used on moorings, but rather on bottom mounted platforms.

Currents can be measured with several alternative methods. Mechanical hand instruments lowered from a small boat can be used with advantage in sheltered areas to obtain current data over relevant limited periods. Profiles are obtained by measuring for an interval, usually 1 to 5 min, at each level. The observations should be combined with salinity and temperature profiles. Such a program can give remarkably good information on the water exchange in coastal areas. For use in the deep sea, refined freefalling instruments have been developed measuring the vertical profile of current, conductivity, and temperature. Some results of such measurements are presented in Chapter 1.

Drogues of various types can also be used to obtain reliable information on the current distribution. Most commonly used are current crosses and window shades of plastic, supplied with suitable weights and floats. It is important that the float is as small as possible so as to minimize its influence on the drifting of the drogue. The current measurements can be carried out from an anchored boat. The float of the drogue is attached to a thin line which is payed out loosely a given length measuring the time required for the drogue to drift that length. Several observations are carried out, combined with wind observations and salinity-temperature profiles. Alternatively freely drifting drogues can be tracked for suitable periods of time. It is possible to track several drogues simultaneously, thus obtaining a picture of the drifting direction and the rate of separation between the drogues. This allows an estimation of the "neighbor" diffusivity using the technique described by Richardson[8] and Stommel.[9] The neighbor concentration $Q(\ell)$ is determined by means of the equation

$$\frac{\partial Q(\ell)}{\partial t} = \frac{\partial}{\partial \ell}\left(F(\ell)\,\frac{\partial Q}{\partial \ell}\right) \tag{1}$$

where ℓ is the separation between two neighboring particles (drogues), called the neighbor separation, t is time, and $F(\ell)$ is the neighbor diffusivity. The function $F(\ell)$ depends upon the separation distance and can thus be made to describe the observations. With $F(\ell) \approx F(\ell_o)$ Stommel[9] finds for $t > 0$

$$Q(\ell_1) = \frac{\text{constant}}{\sqrt{t}}\,\exp\,-\,\frac{(\ell_1 - \ell_0)^2}{4t \cdot F(\ell_0)} \tag{2}$$

The variance is given by

$$\sigma^2 = 2\,t \cdot F(\ell_0) = \overline{(\ell_1 - \ell_0)^2}$$

making it possible to determine $F(\ell_o)$ for the initial separation ℓ_o by observing the separation ℓ_1 at time t. The function $F(\ell)$ is analogous to a diffusion coefficient, and thus the scale dependence of the diffusion can be estimated from a series of drogue experiments covering a range of scales. The drogue technique can with care be extended to depths of about 50 m. Chew and Berberian,[10] for instance, used the technique in the Florida Current. At greater depths recording current meters should preferably be used.

It is important to note that discrete particles are considered in the neighbor diffusivity technique. In most applications we are concerned with the concentration distribution of a diffusing substance. The transformation from the concentration distribution to the neighbor concentration can be determined, but the reverse transformation is more difficult, although possible.[10a]

An important question clearly relates to the transformation between Eulerian and Lagrangian type observations. One possibility is to use Taylor's hypothesis of frozen turbulence being advected past the measuring sensor.[11] The hypothesis implies a relation between the wavenumber k and the frequency ω of a particular component of the motion,

$$\omega = k \cdot U \tag{3}$$

where U is the advection (or mean) velocity. Experiments show that this holds well when the turbulent fluctuations are less than the mean velocity, $u' < U$. This is however not generally true for the conditions in the sea. Despite this, the technique has been used to calculate diffusion characteristics from Eulerian current measurements. Thereby the Hay and Pasquill[12] relation between the Lagrangian (R_L) and the Eulerian (R_E) autocorrelations is used, namely

$$R_E(t) = R_L(\beta\tau) \tag{4}$$

The factor β is usually referred to as the scale factor, expected to be larger than 1. This may be determined from simultaneous observations which make it possible to calculate the autocorrelations. This is however not easy to carry out in the field. Alternatively β may be obtained from an expression of the kind

$$\beta = \frac{U}{\left(\overline{v'^2}\right)^{1/2}} \tag{4a}$$

where U is the mean velocity in the direction of the mean current and $\overline{v'^2}$ is the variance of the transversal velocity component. It is usually found that $\beta > 1$. Current measurements in coastal waters in the Kattegat-Skagerrak area have for instance given $\beta \simeq 1.7$ during both summer and winter periods. Krasnoff and Peskin[13] found values in the range 1.5 to 11.3.

Taylor[11] showed that the horizontal eddy diffusion coefficient for fluctuating velocities smaller than the mean velocity is given by

$$K_h = \overline{u'^2} \int_0^\infty R_L(\tau) \, d\tau \tag{5}$$

By means of Equation 4 we find

$$\int_0^\infty R_E(\tau) \, d\tau = \int_0^\infty R_L(\beta\tau) \, d\tau = \frac{1}{\beta} \int_0^\infty R_L(\beta\tau) \, d(\beta\tau) = \frac{1}{\beta} \int_0^\infty R_L(t') \, dt'$$

and hence

$$K_h = \overline{u'^2} \cdot \beta \cdot \int_0^\infty R_E(\tau) \, d\tau \tag{6}$$

It is noted that the use of Equation 4a in the result given by Callaway[14] implies β^2 instead of β in Equation 6.

The Eulerian autocorrelation

$$R_E(\tau) = \frac{\overline{u'(t) \cdot u'(t + \tau)}}{\overline{u'^2}}$$

can be determined from current measurements by moored instruments. Callaway[14] used the technique and assumed that $\beta = 1$ independent of depth, and inherently that $u' \ll U$. These assumptions are questionable, but the technique allows a determination of the diffusion characteristics at various depths and different regions. It may be expected that the relative distribution of K_h should be reliable. It can also be expected that with $\beta = 1$ a conservative estimate of the dilution would be obtained through the use of the resulting diffusion coefficients.

In many applications it is important to find the diffusion perpendicular to a well-defined mean current. The lateral diffusion will then be related to the Lagrangian autocorrelation function for the lateral current component. The precise nature of the relation between the Eulerian and the Lagrangian correlations is, however, a relatively complicated question.[15]

Information on currents, water exchange, and mixing can also be obtained by means of various tracers. The use of large-scale, long-term average salinity, temperature, and oxygen distributions in combination with various circulation models to determine mixing characteristics for the ocean interior was discussed in Chapter 1. The range of values obtained was surprisingly narrow and suggested that the interior mixing is very weak. Also, in semi-enclosed regions like the Baltic and the Mediterranean, the salinity is a very useful tracer, especially in combination with some other natural substance. In the Baltic-Skagerrak region the distribution of yellow substance from the northern Baltic to the Skagerrak shows a distinct relationship with the salinity distribution.[16,17] This makes it possible to define distinct water masses and identify regions of different mixing characteristics. The suspended matter is another example of a natural tracer which can give very useful information on the mixing characteristics (see, for example, Jerlov[16] and Chapter 1).

In some cases, however, artificial tracers can be used with advantage to study the mixing as well as current distribution and water exchange. Several tracer techniques have been developed and occasionally used in combination. Biological tracers in the form of bacteria from outfalls have also been used for limited scales. In order to interpret the tracer studies environmental observations of the kind discussed earlier have to be included in the field investigations.

III. DYE TRACER TECHNIQUES

A great number of tracers are available. Finnish oceanographers carried out the first tracer study in the Baltic Sea area using fluorescine. It is usually possible to find a suitable tracer for each purpose, with cost and available instrumentation also being worth considering.

It is important to consider the purpose of the tracer experiment, the decay properties of the tracer and its possible interaction with or dependence on the natural conditions such as salinity, suspended matter, and botton fauna, for instance, by adsorption. In many applications information on dilution and residence time in a given region are primary goals. Often it can be advantageous to use parallel injection of two tracers with different decay properties, making it possible to determine mixing rates and residence time[18] (see Chapter 2).

The tracers detected by their fluorescence are relatively cheap, completely harmless in the concentrations normally used in the sea, easily available, and they may be traced with instrumentation that is not too complicated. The ideal dye tracer should fulfill several criteria, such as good stability and good solubility with no precipitation caused by sea salt. The tracer rhodamine (B, C, 5G, BMG) introduced by Pritchard and Carpenter[19] is traced through its fluorescence, in the case of rhodamine B peaked at 575 nm, and excited with green mercury light at 546 nm. This is a wavelength interval situated in a range of weak natural fluorescence by marine plankton and chlorophyll, although Pritchard[20] reported algae with the same spectral properties. The interval is also large enough to make it possible by optical filters to eliminate the primary light scattered by particles in the water.

Rhodamine is a dye of the same chemical group as fluorescine and can be purchased dissolved in acetic acid in a concentration around 30%. The fluorescence is influenced by the temperature, decreasing about 2.5% per °C increase in temperature, but not significantly by the salinity.[21-24] The dyes are differently adsorbed by organic matter, BMG much less than B. Exposed to direct sunlight the fluorescence decreases about 2%/hr for rhodamine, much faster for fluorescine. In the sea the decay will be reduced considerably due to the conditions. The density of the dye solution can be lowered to that of sea water by adding methanol. Density adjustment can also be achieved by using fresh water and glycerine or sugar. Precipitation in sea water can be prevented by diluting the original dye solution well below the dye solubility limit.

The density of the dye solution can be measured with hydrometers which are also used to measure the sea water density. In this way the density of the dye solution can be adjusted with an accuracy of $\pm (0.2 \text{ to } 0.4) \cdot 10^{-3}$. When making subsurface releases it is sometimes appropriate to adjust the temperature of the solution approximately to the level at the depth of injection.[22,25]

The dyes can be detected *in situ* over a large range. In coastal waters normally from about 10^{-12} to 10^{-6} ton m^{-3}. The background signal can vary quite considerably from region to region, and it is normally necessary to carry out a background mapping before doing the actual tracer measurements. Weidemann[22] gives an account of the properties of various dyes as well as the available instrumentation for the tracing. The most commonly used in vitro instrumentation is the Turner® (Turner® Designs) fluorometer first described by Pritchard and Carpenter.[19] Based on long-time experience Carter[23,24] discusses modifications and calibration procedures of that type of instrumentation. It can be concluded that the equipment is very suitable for studies in shallow water and when only surface layer measurements are required in deep water.

For tracing below the near surface layer some sort of *in situ* fluorometer is required. Early attempts to accomplish this involved lowering of an in vitro type instrument in a pressure-tight house and pumping the water past the measuring volume. This was not practical. Essentially four models of real *in situ* fluorometers have been developed, namely by Costin,[26] Karabashev and Ozmidov,[25] Kullenberg[27,28] and Früngel et al.,[29] all discussed by Weidemann.[22] The proper exciting wavelength is selected from the light source by optical glass filters, e.g., Schott® OG2/1 mm, BG18/3 mm, BG20/6 mm. The detector can be a photomultiplier tube or a photodiode, covered with an appropraite filter combination, e.g., Schott® VG9/2 mm and Kodak® Wratten 23 A. The detection range varies between 3 and 5 decades. The instruments are calibrated in the laboratory at suitable intervals.

The basic problem for the *in situ* instruments when operating in the surface layer is the elimination of the ambient daylight. This can be accomplished by using a chopped light source[29,29a] (Figure 1) or by using a second detector which detects only the daylight[27] (Figure 2). The chopped light source may create technical problems whereas the other method cannot completely eliminate the daylight due to differences between the detector units.

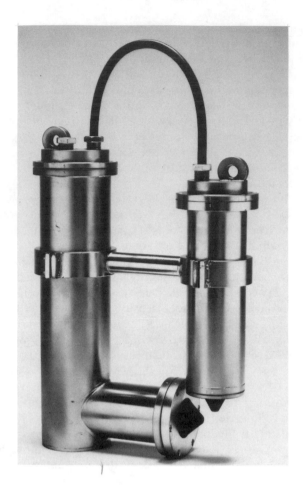

FIGURE 1. In situ fluorometer with a chopped light
source (after 29 a).

Another problem concerns the towing speed and keeping the instrument at the
proper depth. For the case of small-scale experiments a low cruising speed is appropriate and the instrument can be kept at a proper depth without the use of any paravane
device. Such a device is, however, required for any reasonable towing speed. Thereby,
sophisticated equipment such as a batfish can be used, where the fluorometer is attached to the towing body which is vertically cycling within a given interval. Alternatively, a paravane of the triangular wing-formed type can be suspended together with
the fluorometer. Heavy, fish-like weights have also been used together with faired
cables.

In the case of subsurface tracer experiments the tracing is greatly facilitated when
the temperature and preferably also the salinity are measured together with the fluorescence. Normally the dye will become distributed in a fashion similar to the S,T
distributions and hence it can be much more easily followed than if these distributions
(or at least one of them) are not observed simultaneously. Clearly a much more reliable
interpretation of the tracer experiment can also be made in that case.

The injection of the tracer solution can be accomplished in different ways, partly
depending upon whether the aim is a so-called instantaneous release or a continuous
release over an extended period. The latter approach may be preferable when investigating the dispersion in relation to sewage outfalls. When a theoretical model is avail-

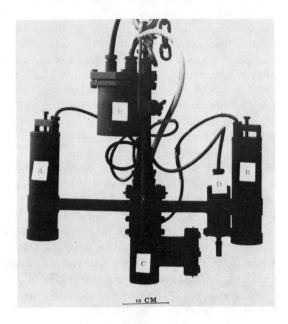

A

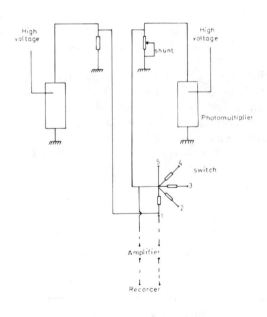

B

FIGURE 2(A). *In situ* fluorometer using two detectors. A and B, photomultipliers; C, lamp house; D, thermistor unit; E, depth indicator. (B) Schematic over connection of detectors in fluorometer shown in Figure 2A.

able for the interpretation, instantaneous releases may be sufficient, which are easier to perform.

Elaborate techniques have been developed for the continuous release (e.g., Murthy and Csanady[30]). Normally it is possible to simulate a scaled outfall quite well both with surface and subsurface (bottom) injection. For the instantaneous surface release Joseph et al.[31] (see also Weidemann[22]) used a device by which the dye solution was forced by compressed air through a hose from the tank onboard to an outlet on a surface buoy at a small distance from the ship, thus avoiding direct disturbance of the ship during the injection. A small volume of dye solution can of course simply be poured into the water directly from the ship. Subsurface injections can be made by forcing the solution through a hose out via a vertical diffusor at the selected depth or by lowering a container with the solution which is opened at the selected depth (see, for example, Kullenberg[28]). In these ways a small initial dye patch is obtained which requires careful navigation for successful tracing.

During the injection, which can last 5 to 10 min, the ship is keeping a position relative to a surface buoy connected to a large drogue at the depth of injection. The positioning in this phase can be done by optical means or by radar. After the injection the tracing starts immediately with the fluorometer being towed across the area of injection navigating relative to the buoy. During the initial phase of the tracing the dye patch and the drogue are normally drifting in a similar fashion. The relative navigation is maintained until the position and drifting direction of the dye patch have been reliably established. Often the dye will become distributed in relatively thin layers and in order to ensure finding of the layer it is necessary to use the fluorometer in a sawtooth fashion over a suitable depth interval, normally 10 to 20 m.

After the initial phase of the tracing an alternative navigation system is preferably used, such as Decca. In some areas this is so accurate that no separate technique is needed initially. A plot should be maintained throughout the tracing. Preferably a systematic cruising fashion should be used so as to cover the dye patch well enough to permit reliable concentration contouring.

A similar navigation approach can be used for surface layer tracing. The initial phase is not so critical in this case since the dye patch will be visible. In the case of a continuous injection a plume is formed downstream which normally will oscillate in position quite markedly. A common approach is to map the concentration distribution in sections perpendicular to the mean plume direction at a series of distances from the point of injection. Each section has to be repeated many (20 to 50) times in order to secure a statistically significant material.

The duration of the tracing period will depend upon the possibility of mapping the dye patch in a reasonable period of time. For a small patch one ship is sufficient and such experiments can last up to several days or a week.

The practical depth limit of these types of tracer experiments is about 100 m. It is clearly very difficult to carry out such experiments at greater depths, although some successful examples exist (see Chapter 1).

The tracing will yeild an in-principle three-dimensional concentration distribution as a function of time. The distribution is normally asymmetrical with sharp concentration variations, especially vertically but also horizontally. For subsurface experiments the depth of the dye layer will often vary, following the salinity and temperature distribution. Despite this the layer is normally assumed to be horizontal and a horizontal concentration distribution is constructed. This can be used to calculate the concentration variances in the direction of the principal axes (σ^2_x and σ^2_y). The problem is often to define these axes. This problem is avoided if the observed distribution is transformed into an equivalent rotationally symmetrical distribution, which can be used to calcualte the symmetrical variance (σ^2_{rh}) and the Joseph-Sendner diffusion velocity

P,[33,33a] as well as other diffusion parameters. The observations in the vertical direction can be used to determine the rate of vertical mixing by applying various models (see Chapters 1 and 2).

Successful tracer experiments will yield a lot of useful information in particular when combined with the relevant environmental ovservations. These should primarily give information on the density and current distributions and their variability. Repeated profiles obtained in conjunction with the tracing are used to determine average gradients of density and current, using an appropriate vertical resolution and and appropriate averaging time period. Richardson numbers and Froude numbers can then be calculated. The wind is another important factor which must be observed at proper intervals. It should be noted that tracer experiments only give information over a fairly limited range of scales.

In order to obtain information on the dispersion during different environmental conditions a series of experiments must be performed. Clearly one should investigate in particular situations which might be of concern. With a large enough material, semi-empirical relations between the mixing and the environmental conditions may be investigated (see, for example, Kullenberg[28]). In particular the influence of the stratification and the wind should be considered.

An alternative tracing method is to use some sort of remote sensing. Surface layer dye patches can be photographed from airplanes and the pictures evaluated by a special technique to yield really synoptic concentration distributions.[33b] A small amount of dye $\approx$ 10 kg) was by this technique traced for about 8 hr in the Baltic. It is noted that no sophisticated instrumentation is required for the field phase and that a considerable amount of information can be obtained in a short time over a large area.

Sattelite imagery and aerial photogrammetry may also be used to obtain information on conditions such as wave characteristics, features of surface circulation, sources and patterns of anomalous water, erosional and depositional features, and dispersion characteristics in general. River water plumes and plumes of heated water effluents can be identified. In the open sea eddies of various kinds and fronts can be identified and traced. Reasonable quantitative information about various parameters can be derived, but at least some sea-truth data are required for a proper interpretation of the imagery. False features can be mistaken for real.

IV. ALTERNATIVE TRACER TECHNIQUES

In many applications it is of interest to use alternative artificial tracers, of which the most important ones are various radioactive isotopes. For local dispersion studies injections of small amounts of some radioactive tracer can be used in very much the same way as the dye tracers. The isotope ^{82}Br (in the form of ammoniumbromid) has become widely used.[34] The counting device is lowered *in situ* behind the ship in the same way as the fluorometer used for the dye technique.

When diluted in a volume of $40 \cdot 10^6$ m^3 water 1 Ci of ^{82}Br can be detected. The use of radioactive tracers must be adjusted to regulations and safety requirements. When a large number of experiments shall be carried out, or in the case of submerged experiments, it appears to be more practical to use dye tracer rather than a radioactive tracer.

Occasionally it is advantageous to use two tracers with different decay times in a so-called parallel injection.[35] Thereby the continuous injection technique is used and the tracers are selected on the basis of well-defined and different decay parameters as well as the requirement that they have different tracing properties. The aim is often to simulate a discharge of a waste product characterized by known decay factors. A combination of the radioactive tracer ^{82}Br with a half-life of 36 hr and decay constant

0.0193 hr^{-1}, and the dye tracer pontacyl brilliant pink with half-life of 460 hr and decay parameter about 0.0015 hr^{-1}, was successfully used by Cedervall and Hansen[36] to determine residence times in the surface layer of a fjord.

Tracers can also be injected together with the waste discharge with the aim of tracing in particular a submerged outfall trapped in a subsurface layer. This technique can be used for tracing injections of different kinds of contaminated waste water, including for instance those from a refinery. In regions where the recipient is stratified this is a very useful way of finding the waste water after the injection since it often becomes distributed in relatively thin layers at various depths (see also Chapter 1). Once the layers have been found water samples can be taken for analysis of the concentration of the contaminant in question.

In many applications concerned with dispersion of the discharge from sewage outfalls a main factor of interest in the decrease of bacterial concentration during the dispersion in the recipient. Harremoes[1] discussed the various methodologies and techniques available for the determination of microbial disappearance in sea water. He emphasized the need for field investigations tailored to the particular conditions. By combining a tracer technique with observations of coliforms the effects of dilution and the combination of dilution and die-away can be determined experimentally. In this way an optimal solution may be obtained to the problem of the site and length of an outfall in relation to water quality requirements and costs. Measurements on continuous sewage and tracer discharge together with current observations and known discharge rates can be used to determine the bacterial disappearance at different distances from the outfall. An abrupt tracer release into a continuous sewage discharge permits a more accurate determination of the drift time and hence the rate of disappearance. These methods constitute the so-called mass balance approach which Harremoes[1] recommends. It is to be noted that there are severe difficulties both in regard to the determination of the dilution and the determination of the bacteria concentration, implying that a large variability in the results can be expected. Harremoes considered that the differences found in the T_{90} values between investigations in different parts of the world were rather due to experimental difficulties than to real T_{90} differences (T_{90} being the drifting time required for a 90% disappearance). Coliform T_{90} values given by Harremoes are in the range 45 to 100 min, but larger values can be encountered. It should be pointed out that for dye experiments Okubo[37] considered mass balances between 50 and 150% as acceptable.

Many applications are concerned with the dispersion at dumping sites. Since the material to be dumped normally contains material in suspension as well as in solution it is often appropriate to carry out an *in situ* experiment tracing the dumped material. A similar technique as the one used for subsurface dye experiments can be used to facilitate the tracing. The suspended matter can be traced by measuring the light scattering or the beam transmission in the red part of the spectrum (650 nm). Suitable *in situ* instrumentation is described by Jerlov.[16] It is clearly necessary to carry out a background mapping prior to the dumping operation. The suspended matter can be measured in the range 30 to 0.1 mgℓ^{-1}. The natural concentration is normally about 0.5 mgℓ^{-1} in coastal waters. Exceptions are found in areas like the Southern Bight of the North Sea where the concentration is much higher.

The dissolved part of the dumped material may be traced by a dissolved tracer (e.g., dye). In many cases the suspended matter is, however, of primary interest. In order to interpret the dumping experiments information is required on the density and general characteristics of the dumped material. The standard environmental observations of density, currents, and wind are also necessary parts of the experiment.

V. MESO- AND LARGE-SCALE EXPERIMENTS

The meso-scale dispersion along coasts and in shelf sea areas can be investigated by means of releases of radioactive material from coastal power plants. Isotopes such as ^{60}Co, ^{58}Co, ^{65}Zn, ^{54}Mn, and ^{137}Cs are used. Measurements can be made on water samples, on phyto- and zooplankton samples, and on suitable samples of fauna as well as sediment samples. In this way a good picture of transportation and distribution of the released material among various compartments as well as over the area is obtained. Kautsky[38] presented several years of observations in the North Sea area of the distribution of ^{137}Cs released from power plants at Cherbourg in the English Channel and Windscale in the Irish Sea. The observations showed transports from the Irish Sea around Scotland to the southern North Sea was well as transports through the English Channel along the eastern North Sea coast into the Skagerrak. Thus considerable information on the water movements and residual current velocity could be obtained.

Becker[39] compared simultaneous salinity and ^{137}Cs measurements and showed that the mixing plays an essential role for the distribution of ^{137}Cs in the North Sea. Mattsson, Finck, and Nilsson[40] applied the technique to the dispersion along the Swedish coast in the southern Kattegat area over a distance of about 100 km. It is quite clear that this is a very powerful tool which can be used in many areas. In this connection the possibility of using river outflows for mixing studies should be mentioned (see also Chapter 1).

A powerful tool for large-scale dispersion studies in the ocean is the use of tritium transferred to the ocean through fallout from the bomb tests in the 1960s.[41-43] The most extensive observations have been carried out in the Atlantic, but there are also observations from the Pacific. The half-life of tritium makes it a very suitable tracer for investigating the diffusion in the top kilometers of the ocean on time scales of years. For the subtropical areas these investigations were the first to demonstrate how weak the internal diffusion is in the ocean. The investigations have also shown that the mixing can be much more intense in high latitudes (see also Chapter 1). The advection of high-latitude water towards subtropical regions in the bottom layer as well as at intermediate ($\approx$ 1000 to 1500 m) depths has also been demonstrated.

The isotopes ^{222}Rn and ^{228}Ra have been used to obtain information on the vertical mixing in the bottom boundary layer of the deep ocean.[44a] These results show a fairly marked vertical mixing in the bottom layer, sometimes extending up to 1500 m from the bottom, at least two orders of magnitude larger than in the main thermocline region. It is quite clear that the use of various radioactive isotopes is a very powerful technique and that many different isotopes can be used once the analytical and sampling tools have become reliable. The distribution of identified pollutants can also be used to make inferences about the dispersion. The case of the spreading of DDT through the atmosphere is well known. Harvey and Steinhauer[44] used the PCB distribution in the Atlantic ocean in the combination with the properties of PCB as well as known features of the circulation to infer possible transfer paths of PCB in the deep ocean.

The mixing conditions can obviously also be investigated by means of the S,T distributions, using both the small-scale structure and the average (large-scale) distributions (see also Chapter 1). The so-called T-S diagram, often using the potential temperature, has been used in many classical investigations (see, for example, Sverdrup et al.,[45] Defant,[46] and Jacobsen[47]) showed that a combination of consecutive T-S profiles plotted in a T-S diagram could be used to determine the ratio between vertical mixing and horizontal velocity, assuming a balance between advection and vertical mixing. The so-called "core" method of tracing the extension of water masses, perhaps more appropriately called water types, is also well known (see, for example, Defant[46]).

Also in the open ocean discrete "particles" can be used to study the dispersion. Specially constructed floats which could be traced acoustically have been used by Freeland et al.[48] These experiments clearly demonstrated the intermittency of the mixing also on large scales. A technique which is being developed makes use of satellite-tracked surface buoys. This was used by Richardson et al.[49] for tracking Gulf Stream rings, demonstrating very nicely the potential of the method for obtaining information on water movements.

VI. CONCLUSIONS

There exists a number of techniques to investigate the dispersion in coastal waters and the surface layer of the sea on small to meso-scales. In order to obtain a proper interpretation of the field studies careful planning of the observations is required so that the appropriate techniques are used and all the relevant observations are included. It is necessary to consider the variability of the environment and avoid limiting the observations to certain conditions only. Field studies must normally be combined with various theoretical models. Often the field studies are carried out for the purpose of obtaining data for testing and calibration of numerical dispersion models.

The most rewarding field investigations in the present context are planned and carried out jointly by interdisciplinary groups. In this way much relevant information may be obtained.

REFERENCES

1. **Harremoes, P.,** In situ methods for determination of microbial disappearance in sea water, in *Discharge of Sewage from Sea Outfalls*, Gameson, A. L. H., Ed., Pergamon Press, Oxford, 1975, 181.
2. **Foxworthy, J. E. and Kneeling, H. R.,** Eddy Diffusion and Bacterial Reduction in Waste Fields in the Ocean, Rep. no. 69-1, Allan Hancock Foundation, University of Southern California, Los Angeles, 1969.
3. **Harremoes, P.,** Diffuser design for discharge to a stratified water, *Water Res.*, 2, 737, 1968.
4. **Larsen, I. and Sørensen, T.,** Buoyancy spread of waste water in coastal regions, in Proc. Coastal Engineering Conf., London, 1968, 1397.
5. **Hydén, H. and Larsen, I.,** Surface spreading, in *Discharge of Sewage from Sea Outfalls*, Gameson, A. L. H., Ed., Pergamon Press, Oxford, 1975, 277.
6. **Hansen, J. A. and Jensen, P.,** Hydrographic and Hydraulic Marine Outfall Design, WHO Training Course on Coastal Pollution Control, Vol. 3, DANIDA, Copenhagen, 1978, 721.
7. **Shaffer, G.,** Conservation calculations in natural coordinates (with an example from the Baltic), *J. Phys. Oceanogr.*, 9, 847, 1979.
7a. **Nielsen, P. B., and Jacobson, T. S.,** An Intercomparison of Acoustic, Electromagnetic and Laser Doppler Current Meters at STARESO 1975, Rept. no. 41, Inst. Fys. Oceanogr., University of Copenhagen, Copenhagen, 1980.
8. **Richardson, L. F.,** Atmospheric diffusion shown on a distance-neighbour graph, *Proc. R. Soc. London Ser. A*, 110, 709, 1926.
9. **Stommel, H.,** Horizontal diffusion due to oceanic turbulence, *J. Mar. Res.*, 8, 199, 1949.
10. **Chew, F. and Berberian, G. A.,** Neighbour diffusivity as related to lateral shear in the Florida Current, *Deep Sea Res.*, 19, 493, 1972.
10a. **Okubo, A.,** Horizontal Diffusion from an Instantaneous Point Source Due to Oceanic Turbulence, Tech. rept. 32, Ref. 62-22. Chesapeake Bay Institute, The Johns Hopkins University, Baltimore, 1962.
11. **Taylor, G. I.,** Diffusion by continuous movements, *Proc. London Math. Soc. Ser. 2*, 20, 196, 1921.
12. **Hay, J. S. and Pasquill, F.,** Diffusion from a continuous source in relation to the spectrum and scale of turbulence, *Adv. Geophys.*, 6, 345, 1959.
13. **Krasnoff, E. and Peskin, R. L.,** The Langevin model for turbulent diffusion, *Geophys. Fluid Dyn.*, 2, 123, 1971.
14. **Callaway, R. J.,** Subsurface horizontal dispersion of pollutants in open coastal waters, in *Discharge of Sewage from Sea Outfalls*, Gameson, A. L. H, Ed., Pergamon Press, Oxford, 1975, 297.

15. **Franz, H. W.**, On Lagrangian and Eulerian correlations, *Rapp. P. V. Reun. Cons. Int. Expl. Mer,* 167, 125, 1974.
16. **Jerlov, N. G.**, *Marine Optics,* Elsevier, Amsterdam, 1976.
17. **Højerslev, N. K.**, Inherent and Apparent Optical Properties of the Baltic, Rept. no. 23, Inst. Fys. Oceanogr., University of Copenhagen, 70 pp.
18. **Cederwall, K., Göransson, C. G., and Svensson, T.**, Subsequent dispersion-methods of measurement, in *Discharge of Sewage from Sea Outfalls,* Gameson, A. L. H., Ed., Pregamon Press, Oxford, 1975, 309.
19. **Pritchard, D. W. and Carpenter, H. H.**, Measurements of turbulent diffusion in estuaries and inshore waters, *Bull. Int. Assoc. Hydrol.,* 20, 37, 1960.
20. **Pritchard, D. W.**, On dye diffusion techniques used in Chesapeake Bay Institute, In Symp. Diffusion in Oceans and Fresh waters, Ichiye, T., Ed., pp. 146—147, Lamont Geological Observatory, Columbia University, Palisades, New York, 1965.
21. **Feuerstein, D. L. and Selleck, R. E.**, Tracers for Dispersion Measurements in Surface Waters, SERL Rept. no. 63-1, Sanitary Engineering Research Laboratory, 1963.
22. **Weidemann, H.**, The use of fluorescent dye for tubulence studies in the sea, in *Optical Aspects of Oceanography,* Ed. Jerlov and Steemann Nielsen, Academic Press, New York, 1974, 257.
23. **Carter, H. H.**, The measurement of rhodamine tracers in natural systems by fluorescence, *Rapp. P. V. Reun. Cons. Int. Explor. Mer,* 167, 193, 1974.
24. **Carter, H. H.**, Prediction of far-field exclusion areas and effects, in *Discharge of Sewage from Sea Outfalls,* Gameson, A. L. H., Ed., Pergamon Press, Oxford, 1975, 363.
25. **Karabashev, G. S. and Ozmidov, R. V.**, *Izv. Acad. Sci. USSR Atmosph. Oceanic Phys.,* 1, 1178, 1965.
26. **Costin, J. M.**, Dye tracer studies on the Bahama Banks, in Symposium on Diffusion in Oceans and Fresh Waters, Ichiye, T., Ed., Lamont Geological Observatory, Columbia University, Palisades, N.Y., 1965, 68.
27. **Kullenberg, G.**, Measurements of Horizontal and Vertical Diffusion in Coastal Waters Rept. no. 3, Inst. Fys. Oceanogr., University of Copenhagen, 1968.
28. **Kullenberg, G.**, An Experimental and Theoretical Investigation of the Turbulent Diffusion in the Upper Layer of the Sea, Rept. no. 25, Inst. Fys. Oceanogr. University of Copenhagen, 1974.
29. **Früngel, F., Knütel, W., and Suarez, J. F.**, Meerestechnik 1, *Mar. Technol.,* 2, 241, 1971.
29a. **Hundahl, H. and Holck, J.**, A New in situ Fluorometer for Rhodamine and Chlorophyll *a* Measurements, in Rept. no. 42, Inst. Fys. Oceanogr., University of Copenhagen, Copenhagen, 1980.
30. **Murthy, C. R. and Csanady, G. T.**, *Water Res.,* 5, 813, 1971.
31. **Joseph, J., Sendner, H., and Weidemann, H.**, Untersuchungen über die horizontale Diffusion in der Nordsee, *Dtsch. Hydrogr. Z.,* 17, 57, 1964.
32. **Weidemann, H., Ed.**, The ICES diffusion experiment RHENO 1965, *Rapp. P. V. Reun. Cons. Int. Explor. Mer,* 163, 1973.
33. **Joseph, J. and Sendner, H.**, Über die horizontale diffusion im Meere, *Dtsch. Hydrogr. Z.,* 11, 49, 1958.
33a. **Joseph, J. and Sendner, H.**, On the spectrum of the mean diffusion velocities in the ocean, *J. Geophys. Res.,* 67(8), 3201, 1962, 3206, 1962.
33b. **Schott, F., Ehlers, M., Hubrich, L., and Quadfasel, D.**, Small-scale diffusion experiments in the Baltic surface-mixed layer under different weather conditions, *Dtsch. Hydrogr. Z.,* 31, 195, 1978.
34. **Hansen, J.**, Tracer Engineering in Coastal Pollution Control, IAEA, SM - 142/37, 1970.
35. **Cederwall, K.**, Hydraulics of Waste Water Disposal, Rept. no. 42, Hydraulics Division, Chalmers University of Technology, Göteborg, 1968.
36. **Cederwall, K. and Hansen, J.**, Tracer studies on dilution and residence time distribution in receiving waters, Water Res. IAWPR, 1967.
37. **Okubo, A.**, Oceanic diffusion diagrams, *Deep Sea Res.,* 18(8), 789, 1971.
38. **Kautsky, H.**, The distribution of the radio nuclide caesium 137 as an indicator for North Sea watermass transport, *Dtsch. Hydrog. Z.,* 26, 241, 1973.
39. **Becker, G.**, Sea surface-water salinities observed during Deutsche Hydrographisches Institut North Sea radiological survey, *Dtsch. Hydrogr. Z.,* 26, 247, 1973.
40. **Mattsson, S., Finck, R., and Nilsson, M.**, Temporal and Spatial Variations in the Distribution of Activation Products from Barsebäck Nuclear Power Plant (Sweden) in the Marine Environment as Established by Seaweed, Environmental Pollution, in press, 1981.
41. **Rooth, C. G. and Östlund, H. G.**, Penetration of tritium into the Atlantic thermocline, *Deep Sea Res.,* 19, 481, 1972.
42. **Östlund, C. G., Dorsey, H. G., and Rooth, C. G.**, GEOSECS North Atlantic radiocarbon and tritium results, *Earth Planet. Sci. Lett.,* 23, 69, 1974.
43. **Peterson, W. H. and Rooth, C. G.**, Formation and exchange of deep water in the Greenland and Norwegian Seas, *Deep Sea Res.,* 23, 273, 1976.

44. Harvey, G. R. and Steinhauer, W. G., Transport pathways of polychlorinated biphenyls in Atlantic water, *J. Mar. Res.*, 34, 561, 1976.

44a. Sarmiento, J. L., Feely, H. W., Moore, W. S., Bainbridge, A. E., and Broecker, W. S., The relationship between vertical eddy diffusion and buoyancy gradient in the deep sea, *Earth Planet. Sci. Lett.*, 32, 357, 1976.

45. Sverdrup, H. U., Johnson, M. W., and Fleming, R. H., *The Oceans, Their Physics, Chemistry and General Biology*, Prentice-Hall, New York, 1942.

46. Defant, F., *Physical Oceanography*, Vol. 1, Pergamon Press, Oxford, 1961.

47. Jacobsen, J. P., Beitrag zur Hydrographie der Dänische Gewässer, *Medd. Komm. Havund. Ser. Hydrogr.*, 1(2), 1913.

48. Freeland, H. J., Rhines, P. B., and Rossby, H. T., Statistical observations of the trajectories of neutrally buoyant floats in the North Atlantic, *J. Mar. Res.*, 33, 383, 1975.

49. Richardson, P. L., Cheney, R. E., and Mantini, L. A., Tracking a Gulf Stream ring with a free drifting surface buoy, *J. Phys. Oceanogr.*, 7, 580, 1977.

Chapter 4

AIR-SEA EXCHANGE OF POLLUTANTS

Michael Waldichuk

TABLE OF CONTENTS

I. INTRODUCTION

The discovery of DDT in organisms in many parts of the world oceans far from apparent direct inputs of this insecticide stimulated a great deal of interest, and eventually research, into the mode of transport of DDT and other ecologically harmful substances to distant areas of the marine environment. It was obvious that the mechanism of global dispersion of some of these pollutants had to be by a more rapid mode than was possible by river runoff and by oceanic current systems and eddy diffusion. The trans-Atlantic atmospheric transport of DDT by the Northeast Trade Wind Systems was first deduced from observations by Risebrough et al.[1] The presence of chlorinated hydrocarbons in the Sargasso Sea atmosphere and surface water has been investigated more intensively recently by Bidleman and Olney,[2] who also examined the long-range atmospheric transport of toxaphene insecticide in the western North Atlantic.[3]

The transport of pollutants via the atmosphere had been well established at least two decades earlier, when radioactive fallout, arising from sea-surface and atmospheric testing of nuclear weapons in the South Pacific during the late 1940s and in the 1950s, was identified in many parts of the world.[4] However, radioactive debris in a nuclear cloud, created by the typical high-altitude mushroom formation following an atmospheric atomic explosion, could be easily visualized as drifting to distant places with stratospheric air currents. Atmospheric transport of something as invisible as DDT vapor seemed less obvious.

The increase of carbon dioxide in the atmosphere, and particularly its possible impact on climate, has been viewed with concern in many quarters for some time.[5-7] The influence of man on the atmosphere, including the introduction of carbon dioxide from burning of fossil fuels and the effects on climate, have been reviewed in a number of forums[8-10] in recent years. The exchange of carbon dioxide between the atmosphere and the sea undoubtedly plays a vital role in buffering the overall effect of carbon dioxide on the atmosphere.

The real concern with atmospheric transport of pollutants and introduction into waters is currently associated with sulfur dioxide emissions from industry in western Europe and eastern North America. In Europe, such emissions are carried by prevailing westerly and southwesterly winds from the industrialized western European countries to the Scandinavian countries, where they are washed out by rain and snow as acidic precipitation. Weakly buffered waters of lakes and rivers have become acidified by the sulfurated precipitation and runoff, disrupting the aquatic ecosystem. Many Norwegian and Swedish lakes have lost their fish populations, among other drastic changes in their freshwater ecosystems.[11,12] Comparable effects have occurred in eastern North America, where lakes downwind from large smelters, fossil-fuel-burning thermal plants, and other sulfur dioxide-emitting industries, have suffered a similar fate.[13-15]

Fortunately, seawater is well buffered, and such marked pH changes from atmospheric input of sulphur dioxide or other acidic and alkaline pollutants are not expected. Nevertheless, marine scientists would do well to take note of the findings of the terrestrial atmospheric and freshwater investigators in connection with the entry of sulfur dioxide and other pollutants from the atmosphere into freshwater systems. The effects of atmospheric pollutants on lakes and rivers can provide a basis for early warning for at least certain types of pollutants and special types of areas (e.g., estuaries) in the sea.

In addition to the downward transfer of substances between the atmosphere and the sea, there is also a significant upward flux for certain natural and man-made substances. Recent studies have shown how anthropogenic materials can be enriched in the surface microlayer and ejected into the atmosphere with bursting bubbles and sea spray. Not only petroleum and chlorinated hydrocarbons are injected into the atmosphere in this way but also metals.[16,17] Pathogens could be introduced into the atmosphere by bursting bubbles on the sea surface where raw sewage or sewage sludge is present[18-20] and the potential exists for the spread of pathogens through the atmosphere.[21]

The importance of the surface microlayer is reflected in the statement of the Panel on Air/Sea Interface Exchange Processes at the Workshop on Pollutant Transfer in the Marine Environment, held in Savannah, Georgia in 1976: "This surface microlayer, although minute in volume compared to the atmosphere or ocean, has an influence on the marine biota far out of proportion to its volume, and represents the interface through which all air/sea interchange must pass."[22] The chemistry of the sea surface microlayer has been thoroughly reviewed by MacIntyre,[23] Liss,[24] and Hunter.[25] Wangersky[26] reviewed in detail the various aspects of the surface film of the ocean as a physical environment. The sampling of the surface microlayer on the sea has presented various problems which have been studied by several investigators.[27-29]

The need for scientific information on certain aspects of interchange of pollutants between the atmosphere and the sea in areas removed from local contamination was the basis for a joint proposal, SEAREX (SEA-AIR EXCHANGE), submitted in early 1977 to the U.S. National Science Foundation Office of the International Decade of Ocean Exploration by a number of U.S. universities and the French Centre des Faibles Radioactivités.[30] This project was approved and is now underway.

It is intended in this review to examine the state of knowledge on air/sea transfer processes, with emphasis on chemical and biological aspects, and to identify some of the substances that enter the sea from the atmosphere and vice versa. The detailed effects of physical factors such as winds, waves, temperature and salinity (density) and turbulence, on the exchange of pollutants between the atmosphere and the sea is beyond the scope of this paper. The reader is referred to such reviews on this subject as given by Dyer.[31]

II. PROCESSES LEADING TO TRANSFER OF SUBSTANCES BETWEEN THE ATMOSPHERE AND THE SEA

There is a variety of physical processes involved in transferring substances from the atmosphere to the sea and vice versa. Some of these are illustrated in Figure 1. The sea-surface microlayer is essentially the boundary between the atmosphere and the sea water, and plays a vital role in the transfer of substances across the air-sea interface. The processes that occur there are especially important to the upward flux of materials, and were described by Garrett and Smagin[33] for air-sea exchange of petroleum hydrocarbons. Some of the processes that convey substances from bulk sea water to the sea surface are examined and reviewed here.

A. Atmospheric Input of Pollutants into the Sea

Windom and Duce[22] discussed in general terms processes responsible for transferring substances from the atmosphere to the sea. Pollutants may be present in the atmosphere in various physical and chemical forms. They may also undergo chemical and physical changes while in the atmosphere because of photochemical reactions, solution or absorption in aqueous droplets and gas-to-particle interconversion. To be transported through the atmosphere they must be capable of entering the atmosphere in (1) gaseous or vapor form, which allows them to be integrated directly into the gaseous phase of the atmosphere; or (2) solid or liquid particulate form (aerosol) light enough to be transported by air currents. In the gaseous or vapor form, pollutants are readily dispersed by turbulence and advected by air currents through the atmosphere from

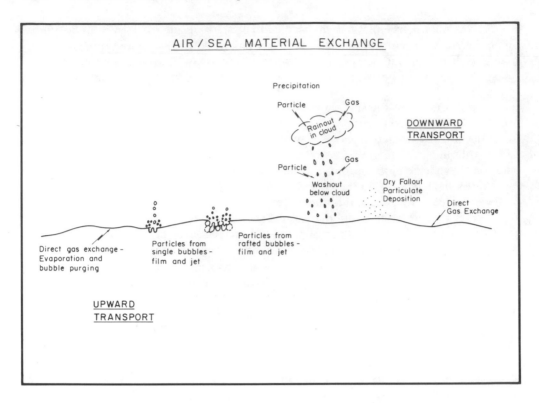

FIGURE 1.   Mechanisms for the exchange of material across the air/sea interface. (From Duce, R. A., *Maritimes*, 22, 4, 1978. With permission.)

place to place without loss by gravitation. Rainout and washout is the primary mechanism for removal from the atmosphere of gases and smaller particles. Through processes of nucleation and condensation in the atmosphere, water droplets form and sorb pollutants that may be present in the surrounding atmosphere. The incorporation of pollutants, gaseous and particulate, into the moisture of clouds and subsequent precipitation is referred to as *rainout*. As the water drops fall in the form of rain, they scavenge further polluting substances from the atmosphere by the process of *washout*, and deposit them on the ground or into water (Figure 1). The precipitation may also come down in the form of snow, and in glaciated areas of the world (e.g., the Antarctic and Greenland icecaps) may provide a rather permanent record of input of substances from the atmosphere over many centuries.[34]

If the substance is a gas at normal atmospheric temperatures (e.g., carbon dioxide, sulfur dioxide, halocarbons and chlorine) then it will be present in the gaseous phase when released to the atmosphere. These atmospheric pollutants have presented no surprises, and their potential seriousness has been recognized for a long time. An appreciable amount of substances with a comparatively low boiling point and high vapor pressure may exist in the atmosphere in vapor form, e.g., water and volatile fractions of petroleum hydrocarbons.

Many of the polluting substances of concern in the atmosphere are liquids or solids at normal room temperature (e.g., DDT, PCBs, and mercury) which have comparatively high vapor pressures. As a consequence, they contribute vapors to the atmosphere when exposed at ordinary atmospheric temperatures. They have built up surprisingly high atmospheric burdens and have been transported over large distances by winds and entered aquatic ecosystems including marine organisms.

The gaseous substances are removed from the atmosphere and deposited in the sea in wet form by incorporation in precipitation by rainout and washout. They may also be injected into the sea in dry form by direct transfer across the air-sea interface.

In the particulate form, the pollutants are maintained in suspension by winds and atmospheric turbulence. Under quiescent conditions, such particulate materials can settle out of the atmosphere by gravitation and form the typical pollutant fallout, which has been a characteristic of radioactive debris introduced into the atmosphere by nuclear weapons tests.[4]

Particulate substances in the atmosphere may be present as solids, either as particles of the pollutant per se, or adsorbed onto finely divided inert matter. When these particles are comparatively large, they are heavier than air and susceptible to settling by gravitation. In the liquid state, atmospheric pollutants occur as tiny droplets, which may be moved through the atmosphere by air currents almost as freely as pollutants present in the gaseous phase. All these tiny atmospheric particles, dry and wet (i.e., anything that can be collected on filter paper), are classed as *aerosols*. Some are produced anthropogenically; others arise from natural sea-surface processes. The production of aerosols from the ocean and their geochemistry have been discussed in considerable detail by Chesselet et al.[35] and MacIntyre.[23]

Air over the sea that has just come off the land will have a particle count usually higher than 10^3 cm^{-3} and often greater than 10^4 cm^{-3}, with most of the particles less than 1 μm in diameter.[36] As the air continues to move over the sea, the particle count decreases with time, presumably as a result of sedimentation, coagulation, and washout by rain. After a few days, a steady state is achieved and the particle count is down to <300 cm^{-3}, which is typical of a "clean" atmosphere.[37] There is some debate whether gas-to-particle conversion produces most of the particles over the sea. It is clear, however, that particles at the large end of the particle spectrum are produced at the sea surface. Such particles, leaving the sea surface as droplets greater than 0.1 μm in diameter, account for 95% of the total mass loading of the atmosphere, normally at 10 to 20 μg cm^{-3}, although they are present in numbers of only 10 to 20 cm^{-3}.[38,39] The particles may undergo a phase change at relative humidities less than 70%, when they become supersaturated with respect to salt and produce "sea-salt particles".[38] One measurement of sea-salt particles in the atmosphere near a surf zone showed a wide range of droplet sizes (<2 to 60 μm diameter) and the total mass of sea-salt was nearly 100 times that normally found in air near the sea surface.[40]

Tropospheric and stratospheric wind systems and atmospheric turbulence are largely responsible for transport, dispersion, and dilution of particulate materials in the atmosphere. Other processes come into play in transferring the pollutants from the atmosphere to the sea. Gravity leads to early deposition of the heaviest particles. Highly pulverized particles of submicrometer size with a large surface area-to-volume ratio are susceptible to movement by atmospheric currents and turbulence, which may keep them in suspension a long time. Particulate matter may be deposited on land and sea in the wet form through rainout and washout by precipitation. It may also be deposited as *dry fallout*. In the dry form it is deposited in the sea by gravitation supplemented by Brownian movement. It may also be trapped by whitecap bubbles.

Dry deposition of atmospheric particles on land or sea surfaces can be evaluated using the concept of *deposition velocity,* v_d, defined by $V_d = 1/M \, dM/dt$ where M is the mass of particles per unit volume (e.g., μg cm^{-3}) and dM/dt is the flux of particles to the surface (e.g., μg cm^{-2} sec^{-1}). The effective deposition velocity varies with particle size, wind speed, and surface roughness; but for particles in the stable aerosol size range, it is often near 1 cm sec^{-1}. A wind tunnel experiment, investigating the particle deposition velocity over a water surface as a function of particle size and wind speed, showed a general decrease in deposition velocity with decreasing particle size at a given wind speed, and a sharp drop in the particle size range of 1 to 10 μm in diameter.[41]

Other studies have shown that dry deposition becomes relatively unimportant for particles smaller than 5 to 10 μm in diameter.[42] They are primarily removed by rainout and washout with precipitation.

Substances vary in their response to atmospheric deposition processes according to their physical and chemical characteristics. Heavy metals, sulfate, radionuclides, aeolian dust, and microorganisms are in the particulate form and are generally deposited by precipitation processes or as dry fallout. Gases, which are water soluble or undergo reaction, are deposited by precipitation or gas-transport processes (air currents and atmospheric turbulence). Nonreactive and nonsoluble gases, e.g., gaseous hydrocarbons and halogenated hydrocarbons, are brought from the atmosphere to the sea surface by gas-transport processes. High molecular weight hydrocarbons and halogenated hydrocarbons can exist in both gaseous and particulate forms in the atmosphere, and may be transferred from the atmosphere to the sea by all four modes. Processes transferring substances from the atmosphere to the sea are summarized in Table 1.

Sampling of atmospheric pollutants to measure air-to-sea fluxes presents problems of contamination. Because of low ambient concentrations of substances to be measured, large volumes must be sampled. To obtain meaningful results, sampling must be conducted at points well removed from anthropogenic sources. Oceanographic vessels are often serious sources of contamination. Rainfall measurements are affected by sea spray. Towers on oceanic islands have been generally the most satisfactory sampling platforms. Bermuda has been utilized for the North Atlantic and Bikini, Enewetak, and American Samoa are serving the SEAREX experiments in the equatorial Pacific.

B. Sea-Surface Microlayer

The sea-surface microlayer is essentially the aqueous air-sea interface. Its thickness varies in response to environmental conditions. According to the technique used to sample it, the microlayer has been described for a thickness ranging from 0.05 to 1000 μm, but it is usually considered to be less than 100 μm thick. Undoubtedly, the chemical composition can vary enormously over this range of thickness. The composition of the microlayer comprises a collection of hydrophobic or surface-active materials, particulates, and microorganisms. Transfer of pollutants through the air-sea interface is effected by processes which are not yet too well understood and in some cases may be mediated by the microlayer.

The surface microlayer, because of its position at the air-sea interface, is exposed to extreme environmental conditions induced by the atmosphere. It constitutes a physical environment quite unlike that of the bulk sea water.[26] The sun may increase light, temperature, and evaporation. On the other hand, sufficient lowering of atmospheric temperature can bring the surface microlayer to freezing or near-freezing conditions. Photochemical oxidation may alter the physical and chemical characteristics of a substance in the microlayer. Ultaviolet (UV) radiation would certainly affect the populations of microorganisms at the sea surface. Substantial temperature increases at the sea surface can arise from intense solar radiation. Evaporation separates the volatile constituents from the nonvolatile residues. Rain may mix the surface microlayer with subsurface sea water and introduce not only freshwater to the sea surface but also substances that are washed out from the atmosphere. Adsorption of certain constituents on floating particulate matter may occur at the sea surface.

Surface winds create waves, which grow with increasing wind intensity, and these may break up or disperse the microlayer. Waves dilate the sea surface by producing a larger surface area for an increase in evaporation and in other sea-to-air transfer processes that depend on exposure of a water surface. Strong wind action coupled with large waves leads to severe disturbance of the microlayer and to spray formation,

Table 1
PROCESSES OF INTERCHANGE OF POLLUTANTS BETWEEN
THE ATMOSPHERE AND THE SEA

Direction of transfer	Processes acting on different forms of pollutant	
	Gaseous and vapor	Aerosol
Atmosphere to the sea	Rainout and washout by precipitation	Rainout and washout by precipitation; deposition by gravitation and Brownian movement (dry fallout)
Air-sea interfacial transfer downward from microlayer to seawater boundary layer	Eddy diffusion (wind mixing); advection; convection; molecular diffusion; emulsification	Trapping by whitecap bubbles; sinking by gravitation; eddy diffusion (wind mixing); horizontal and vertical advection; convection and Brownian movement; molecular diffusion; emulsification
Air-sea interfacial transfer upward from microlayer to atmospheric boundary layer	Evaporation; purging by bubble bursting; escape of gas under supersaturation in water by emission with or without bubble formation	Bubble-bursting; spray by wind shearing of wave crests; bursting of single and rafted bubbles, ejecting jet and film drops into atmosphere
Sea to the atmosphere	Atmospheric turbulence, winds and upward convective air movements	Atmospheric turbulence, winds and upward convective air and Brownian movements

which is perhaps the most important transfer process for nongaseous substances from the sea to the atmosphere. Measurements of spray at an air-water interface have been reported by Wang and Street.[43]

Freshwater inflow in estuaries can cause convergences of surface materials, including the surface microlayer, into concentrations at tidal fronts.[44,45] Substances concentrated vertically in the water column are compressed in the frontal zones by the convergence between the freshwater plumes and the sea water. Thus concentrations of metals, petroleum hydrocarbons, and chlorinated hydrocarbons can be substantially increased at these horizontal interfaces. The surface microlayer, estimated to be 1.5 μm thick, in a Delaware salt marsh was reported to carry, on the average, 10% of the copper, 19% of the zinc, and 23% of the iron relative to the total metal flux of the marsh, including the dissolved and seston components.[46]

Forming a vertical discontinuity, the surface microlayer represents a boundary where certain marine processes start and terminate. It is also a boundary to atmospheric processes. Substances which are too light to remain in the water column, but too heavy and insufficiently volatile to evaporate into the atmosphere, will concentrate at the sea surface. Oil films of natural and man-made origin form part of the surface microlayer in this way.

Any processes that affect the behavior of the sea surface and the composition of the microlayer will influence the air-sea exchange of material. Detergents, petroleum and its derivatives, and the complex mixture of surface-active components from muncipal wastes and sewage sludge alter the air-sea interface and interfacial exchange processes. Petroleum hydrocarbons as well as natural films affect the behavior of the sea surface. Capillary waves are attenuated, and this immediately reduces the sea surface area exposed to evaporation and other transfer processes. Bubble formation may be diminished, and this also reduces the sea-to-air transfer of materials. Although oil films seldom completely seal off the water surface to downward or upward transfer of gases, inasmuch as they are rarely continuous, they may significantly reduce the transfer of gases. Surface-active materials introduced into surface waters may increase frothing and thereby enhance the transfer of substances from the sea surface to the atmosphere.

Surface-active organic materials released by marine plants and animals may alter pollutant exchange processes by modifying interfacial properties with formation of films at the air-sea interface and on air bubbles and particles moving toward this interface. Oleophilic substances, such as petroleum hydrocarbons, chlorinated hydrocarbons, and metallo-organic compounds may accumulate in the films. Sampling the upper 150 μm of sea-surface water in both slick-covered and slick-free areas of the Atlantic and Pacific has revealed the presence of surface-active materials.[27,47,48] Biologically rich areas yielded the greatest quantities of these materials. There is some debate on their composition, with Garrett and co-workers[27,47,48] claiming that they are composed primarily of free fatty acids, fatty esters, fatty alcohols and hydrocarbons. Baier and his associates, on the other hand, suggested that they may be glycoproteins and proteoglycans.[49] More recently, there have been further indications of fatty acids in dissolved and particulate matter in surface films.[50]

The processes of transfer of a substance through the air-sea interface are perhaps most complex and least known in the chain of events in the entry of a pollutant from the atmosphere to the sea and vice versa. Some aspects of the transfer of atmospheric trace constitutents past the air sea interface have been reviewed by Slinn et al.[51] The transfer of gases at natural air-water interfaces has been investigated by Brtko and Kabel.[52] Gaseous substances exchange mainly by molecular processes at the air-sea interface. Gases are mixed and transported by turbulent diffusion and convection processes in the bulk of both the seawater and the atmosphere. Near the air-sea boundary, molecular diffusion dominates in transfer through thin layers on either side of the interface, inasmuch as turbulent diffusion is restricted by the boundary.

Models for direct gas exchange across the air-sea interface[53,54] all assume laminar conditions very close (~ 10 to 100 μm) to the interface and turbulent conditions above and below. They assume that the gases obey Henry's Law, which states that the partial pressure of a gas p in equilibrium with a solution is equal to a constant times its concentration, C, in solution, i.e., $p = HC$. Henry's constant, H, is different for each system and each temperature and must be determined experimentally. For many gases, H is either not known at all, or it has not been measured at the low concentrations involved in ambient air-sea exchange processes. Moreover, accurate measurements have not been made of the low ambient concentrations of most of the exchanging gases above and below the laminar layer. Direct measurements of the air-sea fluxes of gases to validate the models developed for this process are virtually unavailable.

The transfer velocity, V_t (defined as $V_t = D/Z$, where D is the molecular diffusion coefficient from Fick's First Law and Z is the layer thickness) in either the gas phase or the aqueous phase controls the air-sea exchange of a gas. Flux evaluation for gases can be made utilizing the transfer velocity in the rate-controlling phase and the concentration difference driving the interfacial exchange. Such evaluations can be readily made for gases having gas-phase control. However, difficulties are encountered in evaluations for gases with liquid-phase control, because of water solubility and micrometeorological concepts which complicate the calculation of transfer velocities.

Once the atmospheric pollutants deposited on the sea surface break through the interfacial microlayer and boundary layer, then they are subjected to dispersion processes internally in the sea. Advection and diffusion spread the substances horizontally and vertically in the sea water. Reconcentration at the sea surface could only occur through biological processes, where organisms transport the materials into the surface layer by vertical migration, by upwelling processes and by bubble formation, which may concentrate the materials and bring them to the sea surface with the rising bubbles. The known processes of interchange of pollutants through the interfacial layer between the atmosphere and the sea are listed in Table 1.

Sampling of the thin-surface microlayer, without dilution by the subsurface water and contamination by sampling equipment, is critical to satisfactory analysis of constitutents in the microlayer and for an understanding of the processes that occur there. It has challenged many investigators.[27-29,44,46,49,55-65] Table 2 summarizes the advantages and disadvantages of each sampler reviewed. Samplers vary in complexity, thickness of surface layer sampled, contact with subsurface water, portability, and ease of maintenance. Many of the samplers still require a thorough evaluation. An intercomparison of the performance of a number of samplers has been conducted[57,66,67] to broaden the knowledge of suitability of various samplers for specific research objectives, but more such research is needed.

C. Marine Input of Substances into the Atmosphere

The entry of sea salt constituents, such as chloride and sulfate, from the sea into the atmosphere is well known. The atmosphere over coastal land, particularly on the windward side, is often saturated with natural sea salt. There is no reason why anthropogenic constituents might not enter the atmosphere from the sea in the same way.

The upward transfer of gaseous substances from the sea occurs mainly by (1) molecular evaporation from the surface and (2) purging by bubbles. Particulate materials are ejected from the sea into the atmosphere from single bursting bubbles and bubble rafts in the form of jet droplets and film droplets. Such activity as fish and wildlife breaking the surface contribute to some flux of material from the sea to the atmosphere; transfer by this mode has not been estimated. Unusual phenomena such as water spouts, may inject substances from the sea to the atmosphere, but this contribution is probably incidental on a global scale. Atmospheric particles stemming from the ocean are produced primarily by bursting bubbles created by breaking waves. Laboratory wind-wave tank studies, with simultaneous measurements of water droplets in the air and air bubbles in water, have been reported recently.[68] It was concluded that the major source of spray is bursting bubbles rather than wind shearing of wave crests.

When a bubble bursts at the sea surface it produces a jet which ejects 2 to 5 droplets into the air. The jet droplet diameter is about 10% of the bubble from which it was formed.[69] Smaller droplets are also formed by the shattering bubble film cap[70] (Figure 2).

A large proportion of sea-derived particles in the atmosphere is introduced by white caps, which develop at wind speeds of 3 to 4 m sec^{-1}. Since white cap coverage increases nearly as the square of the wind speed[69,71] the flux of liquid particles rises dramatically with increasing winds. The flux of particles to the atmosphere in white cap regions is of the order of 10^3 cm^{-2} sec^{-1} with the majority less than 1 μm in diameter.[72,73] It has been estimated that on a global basis 3 to 4% of the ocean surface is covered by white caps at any one time, and that the overall oceanic production rate of atmospheric particles is approximately 0.1 jet droplets cm^{-2} sec^{-1} and 0.07 film droplets cm^{-2} sec^{-1}. These estimations are based on the bursting of individual bubbles at the sea surface. An important source of these bubble-generated atmospheric particles is also the bursting of bubble clusters or foams, but there is little information on the quantity produced by such clusters.

The formation of bubbles and sea foam through the action of wind and waves on surface-active material in the water facilitates the transfer of substances from the sea to the atmosphere. The significance of bubble formation in the sea to transfer of materials from the sea to the atmosphere has been investigated by Blanchard and co-

Table 2

COMPARISON OF SURFACE MICROLAYER SAMPLERS

Type	Thickness of surface layer sampled (μm)	Description of device and mechanism of sampling	Advantages	Disadvantages	Ref.
Wire screen	100—400	Window screen (16 wires per 2.54 cm) has optimum mesh size screen brought into contact with water surface, withdrawn and drained into appropriate collector	Thickness of surface layer sampled is function of wire diameter; sampler provides a sample of known surface area and thickness; simple to construct, operate, and clean; can be used for natural and pollutant surface films, particulate organic and inorganic materials, and marine organisms	Requires care to prevent collection of floating debris, petroleum residues, jellyfish and sea weed	27
Rotating drum	60—100	Hydrophilic, motor-driven cylinder is rotated at a surface speed slightly less than that of small boat on which sampler mounted; thin film of water lifted upward from the surface; blade pressing tightly against rotating surface at top of cylinder directs collected water into a cup	Samples large areas of water surface without manual dipping; capable of relatively good control of surface layer sampling thickness	More complex than manual dipping methods; requires mounting on a vessel with power; suitable only for relatively calm conditions, usually in coastal waters; restricted portability	28
Glass plate	60—100	Consists of clean, wettable glass plate and a wiper blade made of neoprene or other flexible, non-contaminating material; it is inserted vertically through air-water interface, withdrawn slowly, and water wetting both sides of plate is directed into collection plate with wiper blade	Simple in design and use; highly portable; requires no specialized vessel facilities	May receive extensive subsurface contact and suffer from selective adsorption; collecting characteristics have not been sufficiently evaluated	55
Teflon® plate	50—100	Flat surface of teflon plate is touched lightly to the water surface; plate is then held vertically while adsorbed material is washed off with a stream of solvent into a container	Simple to construct and use; known surface area is sampled; samples closely to surface; good collection efficiency for fatty acids; suitable for routine sampling	Thickness is not known precisely; comparatively low collection efficiency for alkanes and aromatic hydrocarbons	56, 57

Method	Thickness (μm)	Description	Comments	Ref.
Teflon® plate perforated with conical holes	50—100	Perforated Teflon® plate is used with conical holes opening downward toward the water surface	Area of surface sampled is known; monomolecular layers of pure fatty acids and esters have been transferred from the sea surface to the plate at from 70 to 100% area efficiency, when the perforated Teflon® plate was dipped horizontally through the water surface	58
Hydrophilic Teflon® sheet	10—50	Sheet of Teflon® is made hydrophilic by etching with solution of sodium in liquid ammonia; attached to a framework with detachable handles; adsorbs surface film constituents from air-water interface by touching hydrophilic Teflon® sheet parallel to sea surface	Samples known surface area close to the surface; contact reduced with subsurface water; recovers monolayers of fatty acids, fatty esters, and nonpolar alkanes efficiently from water; continues to collect surface film on repeated contact with a monolayer-covered surface	59
Internal reflection prism	0.01—1	Wettable prism dipped through air-sea interface; surface film as thin as a monomolecular layer adsorbed from air-sea interface onto optically polished prisms of suitable material (e.g., germanium); film coatings analyzed in internal mirror assembly which directs IR beam into beveled end of coated prism	Ultrathin surface layers can be sampled and analyzed; multilayers of organic materials (e.g., petroleum films) can also be transferred in suitable condition for IR analyses	49, 60, 61
Bubble interfacial microlayer sampler (BIMS)	0.05—10	Bursting-bubble technique used as microtome to skim off thin layer of sea surface and inject it into atmosphere as jet and film droplets; BIMS suspended between twin hulls of 4-m catamaran, and bubbles of 500—1000 μm diameter produced at variable depths to 50 cm; high-volume air sampling pump draws material injected into enclosed atmosphere in truncated pyramid and deposits it on filters	Sophisticated nonroutine sampler; does not necessarily exclusively collect material in surface microlayer, but also picks up substances scavenged by bubbles rising through water column; limited to wave heights of 0.5 m or less; system cumbersome and difficult to transport	29, 62

Table 2 (continued)
COMPARISON OF SURFACE MICROLAYER SAMPLERS

Type	Thickness of surface layer sampled (μm)	Description of device and mechanism of sampling	Advantages	Disadvantages	Ref.
Polyethylene funnel	50—100	Stoppered polyethylene funnel pushed neck first through sea surface and allowed to fill with subsurface water at least 15 cm below surface film, then moved sideways to undisturbed area and brought slowly to surface entrapping area of film equal to open end of funnel; surface film adhering to inside of funnel is removed with solvent for analysis	Inexpensive; simple; easily transportable	Area sampled only as large as mouth of funnel; tedious procedure; selective adsorption could possibly bias results to more strongly adsorbed species	63
Sterile nuclepore membranes	100—500	Used as adsorbers for slicks and microbes in surface layer; membranes floated on sea surface and retrieved with sterile plastic dishes submersed under them and raised upwards through water surface	Comparatively simple. Thickness of sampled layer not known precisely, but probably greater than with other techniques; gives bacterial and fungal populations several orders of magnitude higher than values previously reported for surface slicks	For microorganisms must be handled aseptically; unsuitable for rough seas	64
Freezing probe	1—1000	Probe consists of disc of 0.25 m diameter polymethyl methacrylate, encased on lower side by thin (<0.1 mm) PVC membrane; probe immersed for 2 min in liquid nitrogen and then touched to surface of sea; surface microlayer freezes to surface of PVC membrane in less than 1 sec	Efficient removal of known area of surface layer; frozen sample can be placed in cryogenic source system of various instruments (e.g., spark source mass spectrometer or scanning electron microprobe microscope coupled to an elemental analyzer); can be used on rough sea	Cumbersome, non-routine procedure; thickness of sea surface sampled not too well controlled; picks up particulate material as well as film	65
PVC film	10—100	Solution of 5% PVC in cyclohexane with a suitable stabilizer (e.g., 1% octylphthalate) sprayed on sea surface; solvent phase evaporates with surface materials trapped in a thin	Efficient recovery of surface material; excludes subsurface water; allows analysis of different parts of sea surface covered by film by cutting out sections of choice	Solvent must be optimal for ambient air and sea surface temperatures; at temperatures below 4°C PVC mixture does not spread rapidly; wave action can compress and distort	65

	Thickness (μm)	Procedure	Advantages	Disadvantages	Ref.
		flexible film of PVC to be removed by wire ring attached to pole; analysis can be done on solid film or on solution by dissolving film in hexane or other appropriate solvent		PVC film at periphery; thickness of surface sampled not well known	
V-shaped PVC boom	1—50 1.5 (average)	Floating V-shaped plastic tube, sealed at both ends and washed with dilute nitric acid, distilled water, methanol, and chloroform; collections made on windward side of ship; material in microlayer collects in apex of V-shaped boom; accumulated aggregates transferred with spatula or plastic screen to a glass beaker	Rapidly collapses hydrophobic and particulate material associated with the microlayer at the apex of the V-shaped boom in convenient, water-free, gram-size quantities; useful in frontal areas where microlayer is compressed laterally	Difficult to quantify thickness and area sampled; unsuited for thinly dispersed microlayer	44, 46

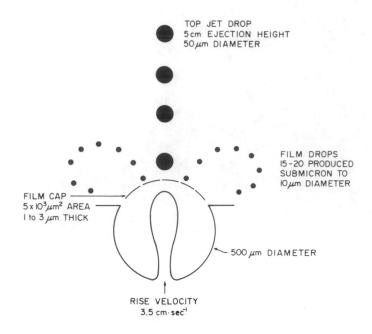

TOP JET DROP
5cm EJECTION HEIGHT
50 μm DIAMETER

FILM DROPS
15-20 PRODUCED
SUBMICRON TO
10 μm DIAMETER

FILM CAP
5 x 10³μm² AREA
1 to 3 μm THICK

500 μm DIAMETER

RISE VELOCITY
3.5 cm·sec⁻¹

FIGURE 2.   Some dimensions of a bubble and of products on bursting.
(Redrawn from MacIntyre.[70])

workers for many years.[37,40,69,73-86] McIntyre recognized the complexity of the mechanism of bubble breaking, especially its hydrodynamics.[87,88] A bubble film may be enriched through selective absorption with a particular material and thus may introduce into the atmosphere on bursting a higher concentration of the material in the bubble droplets than present in the microlayer. Moreover, bubble films can be enriched with certain substances as they rise in the water column. Thus, the particles ejected into the atmosphere may have a substance many times more concentrated than present in the bulk seawater.

When a bubble breaks it skims off a thin-surface layer in a "microtome" effect to form the jet and film droplets. Studies on breaking bubbles have shown that the material in the top jet drop was originally spread over the interior surface of both the bubble cap and the portion of the bubble submerged in a thickness of about 0.05% of the bubble diameter.[69] Therefore, the smaller the bubble diameter the thinner the layer of water surface skimmed off. For bubble diameters of 100 to 1000 μm, which are common in the ocean, the jet droplets are composed of material skimmed off the sea surface to a depth of 0.05 to 0.5 μm. The relationships developed by MacIntyre[23] among jet drop diameter, bubble diameter, and microtome depth are shown in Figure 3. It should be stressed here, however, that bubbles at the sea surface may have also scavenged material from the water column, if they were generated below the surface.

Other processes are also involved in the transfer of substances from the sea surface to the atmosphere, but are probably less effective than wind spray and bubble bursting. One process that may be quite significant at times and in certain areas is rain-drop splashing. A heavy rainstorm associated with a strong wind may substantially enhance the sea-to-air transfer of materials. This was investigated in a preliminary way in early studies by Blanchard and Woodcock,[74] and more recently by Green and Houk.[89] A moderate rain or snowfall on the sea surface can produce as many bubbles per unit area per second as white caps. Precipitation, however, covers the whole area, while white caps do not. Therefore, the average production of bursting bubbles by rain over

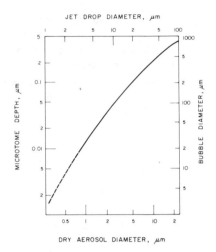

FIGURE 3. Relationship among bubble diameter, jet drop diameter, and microtome depth. (From MacIntyre, F., in *The Sea*, Vol. 5, Copyright 1974 John Wiley & Sons, Inc. Reprinted by permission.)

a large area may be 30 times that produced normally by white caps. Moreover, bubbles produced by precipitation are generally less than 100 μm in diameter and of a size that will remain airborne for a long period of time.

Bubbles can form on the sea surface without wind agitation. Oxygen supersaturation may result from vigorous photosynthesis, and bubbles of oxygen will rise to the surface. Methane may be produced in marine sediments from decomposition of organic matter and gas bubbles will ascend to the sea surface. No estimates have been made of the numbers and sizes of bubbles arising from such biological activity. Marine plants may contribute certain substances other than oxygen to the atmosphere directly if exposed, or through the water if submerged. Algae are known to produce short-chain halogenated hydrocarbons, which exhibit a net flux from the sea to the atmosphere.[54] The release into the atmosphere of particles containing metals by terrestrial vegetation has been documented.[90] Gaseous emissions of mercury from an aquatic vascular plant have also been reported.[91] Emission of biogenic hydrogen sulfide is generally well recognized, and attempts have been made to quantify it.[92] A biogenic source has been suggested as the agent responsible for approximately a 20,000-fold enrichment of copper during aerosol production from the ocean near Tasmania,[93] although this hypothesis could be debated.

A recent concern raised in connection with sea-to-air transfer of substances is the possible transmission via the atmospheric route of pathogenic organisms from sewage and sewage sludge dumped into the sea.[18,94] It has been known for a long time that microorganisms are present in marine air.[95] The irritating effect on the human respiratory system from marine organisms or their metabolic products in the marine atmosphere associated with a massive dinoflagellate bloom in a Florida "Red Tide" was reported by Woodcock.[96] Recent experiments have demonstrated that viruses can be enriched in the surface microlayer and ejected into the atmosphere by bursting bubbles from surf to wind.[18,19,76] A surf deliberately injected with viruses produced an aerosol with 200 times greater virus concentration than the surf itself.[20] The postulated concentrating mechanism is that viruses are adsorbed on air bubbles as they rise through the water column, and when the bubbles burst at the surface, the bubble skin strips the virus-rich layer of water from the bubble surface and ejects it into the air as small droplets.

There is evidence now that some disease-producing microorganisms are transmitted through the atmosphere.[21,97] It has been suggested that certain microorganisms might be encapsulated in oleophilic materials produced by marine organisms at the sea surface and thus preserved for transmission by the atmosphere for long distances. Surface-active material from slicks could form a coating on airborne droplets which, when inhaled, could interfere with the normal function of lung surfactant and lead to respiratory problems and influence phagocytosis.[94]

Film-covered waters have caused destruction of coastal vegetation. The Viareggio phenomenon along the west coast of Italy, bordering the Ligurian Sea, where pines were destroyed by atmospheric pollution, arose from municipal and industrial wastes discharged by rivers.[98] Onshore winds transported petroleum and surface-active constituents from surface films ejected into the atmosphere by bursting bubbles. These substances were coated on the pine needles, thereby interfering with transpiration and resulting in mortality of thousands of trees.

III. POLLUTANTS INVOLVED IN INTERCHANGE BETWEEN THE ATMOSPHERE AND THE SEA

There are many anthropogenic substances that could be interchanged between the atmosphere and the sea. These may be conveniently classified in a number of categories: (1) halogenated hydrocarbons, (2) petroleum hydrocarbons, (3) metals, (4) radionuclides, (5) gases, and (6) microorganisms.

Some of the substances for which the oceans are known to be a sink are DDT, polychlorinated biphenyls, chlorofluoromethane, carbon tetrachloride, and sulfur dioxide. Substances for which the oceans are a source include nitrous oxide, carbon monoxide, dimethyl sulfide, methane, and methyl iodide. Many of the substances with an upward flux between the sea and the atmosphere are believed to stem from biogenic sources.

A. Halogenated Hydrocarbons

A group of compounds that includes such well-known persistent substances as DDT and the polychlorinated biphenyls, the halogenated hydrocarbons are synthetic chemicals most of which are not found in nature and some of which, when introduced by man, can contribute to rather serious ecological disturbances under particular circumstances. Because of their relative stability, some of these chemicals find a variety of industrial uses, the PCB's having been used for many years in electrical transformers and condensers. The short-chain chlorofluorocarbons have found application in aerosol dispensers. Consequently, such gaseous halocarbons as the Freons, used in numerous household pressurized dispensers, have ended up in the atmosphere. The possibility that they may cause erosion of the protective ozone layer in the stratosphere has been demonstrated.[99,100] Their impact, if any, on the oceans has not been fully assessed. However, estimates[54] for trichlorofluoromethane (Freon-11, CCl_3F) show that the oceans are a sink for about 0.5 to 1% of the atmospheric burden of this halocarbon, which is rather similar to the extent (1%) of the stratospheric sink.

Other halogenated hydrocarbons are now recognized as potential hazards to the environment. These include hexachlorobenzene, a grain fungicide, and low molecular weight chlorinated aliphatics, such as trichloroethane, perchloroethylene, and trichloroethylene used as dry-cleaning fluids, solvents and degreasers. All have been detected in air and natural waters. They are relatively stable in the environment, and their residence times in the atmosphere appear to be in the order of a decade. Trichlorofluoromethane, dichlorodifluoromethane, carbon tetrachloride, chloroform, and methyl iodide have all been detected in seawater, and it is believed that most of these reach the ocean via the atmosphere, although the latter three may also be produced naturally and the net flux of methyl iodide is definitely from sea to air.[54]

The sampling and analytical techniques for measuring halogenated hydrocarbons in both the atmosphere and the sea have reached a high degree of sensitivity, precision, and reliability with development of the gas-liquid chromatograph. Measurement of the high molecular weight chlorinated hydrocarbons in air and seawater has been perfected to remarkable reproducibility.[101] With advances in extraction techniques for removal of these substances from large volumes of air and seawater, the sensitivity and limits of detection by existing methods steadily improve.

Using various simplifying assumptions, the net global flux of two high molecular weight halogenated hydrocarbons (PCBs and DDT) and three low molecular weight halogenated compounds (CCl_4, Freon 11, and CH_3I) have been calculated by various investigators[54,101-103] and are shown in Table 3. Except for CH_3I, all the halogenated compounds noted here have mainly an anthropogenic source and the sea is a sink.

B. Petroleum Hydrocarbons

Much concern has been expressed about petroleum hydrocarbons because of the large volume (approximately 6 million tons[106]) that enters the sea annually by deliberate discharge, seepage, accidental small releases, and catastrophic tanker spills. Natural seepage alone has been variously estimated from several orders less than anthropogenic input[107] to a range between 0.2×10^6 and 6.0×10^6 tons annually, with the best estimate for the present seepage worldwide in the order of 0.6×10^6 tons per year.[106,108] A recent report on a finding of a subsurface oil-rich layer in the open North Atlantic ocean suggests that the quantity of oil entering there may be higher than the foregoing best global estimate.[109]

The large-scale utilization of petroleum hydrocarbons by man for his various energy requirements creates a great demand for this fossil fuel. Areas of high consumption are usually far removed from areas of production so that oil must be transported over large distances, often by sea. Sea transport of oil is still the most economical mode for moving oil from areas of production to the major markets of the world. The hazards of sea transport, especially under adverse weather conditions, are well known. Thus, even with extraordinary precautions, spillage in the marine environment is inevitable. Blowouts in offshore oil production, such as occurred in the North Sea Ekofisk oilfield in April 1977 and in the Gulf of Mexico with the IXTOC I oil well in June 1979, can contribute oil to the sea. Oil spills are exposed to the atmosphere both on land and on the sea leading to evaporation, which introduces the low-boiling fractions into the atmosphere. So far, the major impact of oil spills has been largely an economic one with often enormous costs of cleanup. This has been particularly true in areas of high amenity values. The long-term ecological effects of marine oil spills are still not fully understood.

A problem of intercomparison of data from measurements of petroleum contamination in marine samples taken from different areas has been the wide variety of methods used to determine their oil content. The more sophisticated laboratories carry out analyses with mass spectrometry/gas chromatography systems coupled to dedicated computers. Other laboratories may use the simpler fluorescence spectrophotometry technique. The two approaches actually measure different things. Intercalibration exercises have been conducted to determine the accuracy and precision of hydrocarbon analyses carried out by particular techniques. However, there is still no truly routine, universally accepted standard method for determination of petroleum hydrocarbon levels in seawater, marine sediments, and biota. Moreover, there is still no simple way

Table 3
ESTIMATED NET GLOBAL FLUX OF SOME SUBSTANCES BETWEEN THE ATMOSPHERE AND THE OCEANS

Substance	Downward flux (atmosphere to oceans) g yr^{-1}	Upward flux (oceans to atmosphere) g yr^{-1}	Ref.
Sulfur dioxide (SO$_2$)	10^{13}		53, 105
Carbon tetrachloride (CCl$_4$)	1.4 × 10^{10}		53, 103
Chlorofluoromethane (Freon-11, CFCl$_3$)	5.4 × 10^{9}		53
Polychlorinated biphenyls (PCBs)	2 × 10^{9}		102, 103
DDT	2 × 10^{8}		102, 104
Nitrous oxide (N$_2$O)		1.2 × 10^{14}	53
Carbon monoxide (CO)		4.3 × 10^{13}	53
Dimethyl sulfide [(CH$_3$)$_2$S]		7.2 × 10^{12}	53
Methane (CH$_4$)		3.2 × 10^{12}	53
Methyl iodide (CH$_3$I)		2.7 × 10^{11}	53

of distinguishing between biogenic and petroleum hydrocarbons. Considering that there is a large natural contribution of hydrocarbons to the sea from biogenic marine sources and to the atmosphere from terrestrial vegetation, the frequent lack of adequate collecting techniques and of routine diagnostic methodology to distinguish these sources continues to hamper progress in this field.

It has been estimated that the annual world-wide anthropogenic emissions of all types of hydrocarbons to the atmosphere result in a non methane hydrocarbon production of 68 MT, (1 MT = 10^{12} g) and that the annual global atmospheric input of petroleum hydrocarbons to the seas is approximately 0.6 MT.[9] Using an average particulate hydrocarbon concentration of about 10^{-9} g m^{-3} in the marine atmosphere, based on preliminary measurements in air samples taken aboard a ship in the open North Atlantic ocean, Duce[110] estimated a global atmospheric value of 1.4 MT for the heavier particulate hydrocarbons (n−C$_{14}$ to n−C$_{33}$). Garrett and Smagin[33] estimated that 1.35 MT are emitted to the atmosphere by natural seeps, transportation sources, and offshore production facilities. These hydrocarbons consist mainly of volatile, gas-phase compounds that are highly paraffinic and stable, and thus not likely to be rapidly converted into particles that would be eventually redeposited in the sea. Estimates of rapid conversion of nonmethane gaseous hydrocarbons to particles in the atmosphere range from 3 to 30 MT per year, while estimates of the remaining "long lived" gaseous hydrocarbons from anthropogenic atmospheric emissions range from 30 to 65 MT per year of which about 90% are produced in the northern hemisphere.[106]

The Working Group on Petroleum and Related Natural Hydrocarbons at the 1975 Miami Workshop on Tropospheric Transport of Pollutants to the Ocean[9] concluded that: ". . . in addition to continental inputs, it seems likely that there is a cycle of petroleum hydrocarbons between the oceans and the marine atmosphere in which (a) there is a net flux of volatile, low-molecular-weight compounds from sea to air; (b) the less stable compounds undergo gas-to-particle conversion in the atmosphere; and (c) the particulate forms are returned to the sea by precipitation processes and as dry fallout."

Duce[111] reviewed available data on the global source, distribution, and fluxes of nonmethane organic matter in the atmosphere. He estimated that the global primary anthropogenic input of particulate organic carbon is about 30 MT yr^{-1}, of which 16 MT is present on particles with a diameter less than 1 μm and concluded that the annual input of organic carbon on large particles (> 1 μm) in the global atmosphere (36 MT) can be explained by primary emissions from anthropogenic sources (14 MT) and such natural sources as the ocean, crustal weathering, and naturally ignited forest fires (22 MT). Of the total particulate organic carbon ($\sim$ 56 MT yr^{-1}) injected into the atmosphere, about two thirds (36 MT) is present on particles with d > 1 μm and the remainder (20 MT) is on smaller particles. The pollution and natural sources, however, cannot account for the global tropospheric burden of small-particle (< 1 μm) organic carbon. To balance the cycle, it was suggested that 80 to 160 MT yr^{-1} of small-particle carbon are required from another source, such as leaves of vegetation and gas-to-particle conversion of natural and anthropogenic organic compounds.

C. Metals

The oceans contain virtually all the known metals, some in higher concentrations than others, depending on the imput and their solubility. As part of the natural geochemical cycle, metals enter the sea from terrestrial sources and atmospheric transport. Activities of man have altered the geochemical cycle in some respects.[112] Metals can be neither created nor destroyed; only their physical and chemical form can be altered. The latter is extremely important in terms of biological availability for uptake by marine organisms and physiological action of the metal on them and on humans. For example, the methylated form of mercury is far more toxic and more readily bioaccumulated than the inorganic form and was eventually implicated in the tragic Minamata disease.[113]

Most metals can readily undergo interchange between the atmosphere and the sea because they can be introduced into the atmosphere in a highly pulverized particulate form, and depending on their vapor pressure, in the gaseous state. At the present time, however, the data are so sparse that it is impossible to make a reliable estimate of the global flux of metals from land to sea and vice versa. Moreover, there is still no simple way of distinguishing anthropogenic sources of metals from natural input. Some success has been achieved in the use of the isotopic ratio of Pb in urban and near-urban areas to identify the component of atmospheric Pb due to combustion of leaded fuel. The Working Group on Metals at the Miami Workshop of 1975 pointed out that there is a great need for data on metals in the atmosphere of the southern oceans, which could provide at least part of the answer on the anthropogenic input of metals into the global atmosphere.[9] Because only 10% of the particulate pollutant sources are located in the southern hemisphere, data from the southern oceans would provide a better basis for assessing anthropogenic impacts.

The major anthropogenic sources of atmospheric metals include the combustion of fossil fuels, the incineration of wastes, the emissions from cement plants, contributions from smelters and surface mining operations, and other industrial sources. There are other activities which undoubtedly contribute a certain amount of metals into the atmosphere, such as excavations for building construction, gravel washing operations, road building, and manufacturing processes of many kinds. Certain individual man-made sources contribute particular metals which in some instances may be significant. These include high-lead contributions from combustion of lead alkyls in gasoline, arsenic among other metals and metalloids from smelters, cadmium from incinerators, and vanadium from residual oils combustion. High-temperature combustion processes release many metals to the atmosphere, and in this case, these metals are associated with submicrometer particulate matter.

Among natural sources of metals, crustal weathering contributes a significant amount of mineral matter to the atmosphere. Natural processes of crustal weathering arise from effects of runoff, freezing, and abrasion by wind. It has been postulated that low-temperature vaporization of some metals or their compounds occurs from crustal rocks,[114,115] although the global importance of this has been questioned.[116] A suggested major source of atmospheric mercury is degassing of the earth's crust.[117] This process can be enhanced by man's activities, such as tilling of the soil and surface mining.

Volcanic activity can contribute metals to the atmosphere intermittently through eruptions. It has been proposed recently[118] that yields of particulate material from volcanoes can be quite large. Metals with a high enrichment factor in the crust also have a high enrichment factor in particles collected in the fumaroles and vents of active volcanoes.[17,118] The enrichment factor for the crust is defined by

$$EF \text{ crust} = \left(\frac{X}{Al}\right)_{air} \Big/ \left(\frac{X}{Al}\right)_{crust}$$

where X/Al_{air} and X/Al_{crust} are the ratios of the metal X to that of aluminum in air and the earth's crust, respectively.

The various physical enrichment processes that occur at the surface of the sea, coupled with mechanisms of injecting the surface seawater into the atmosphere, could contribute to a net flux of metals from the sea to the atmosphere in certain areas.[93] It has been demonstrated by a number of investigators that many metals are concentrated in the surface microlayer of the sea.[120-123] However, there has been some disagreement on the scale of enrichment of certain metals in the sea surface microlayer and on the geochemical significance of such enrichment. Iron, zinc, cadmium, and copper have been shown to be enriched by a factor of several hundred on atmospheric sea-salt particles produced by bubble-bursting in coastal waters.[17,124] Similar studies conducted recently in the North Sea have shown the presence of trace elements, including lead, on droplets produced from bursting bubbles artificially produced beneath the sea surface, to be about 100 times higher than that in bulk sea water.[181] Clearly, this could be a significant source of metals in the marine atmosphere.

Metals can be mobilized by biological activity from sediments into the water and by various physical processes from the sea into the atmosphere. Many metals are now known to be methylated by microorganisms and this renders them more volatile, with possible evaporation into the atmosphere. The highly toxic methylated forms of mercury have been identified with Minamata disease,[113] and the biological methylation of mercury was subsequently demonstrated.[125] Methylated forms of selenium and arsenic are known to be produced by microorganisms in the marine environment, and this has been suggested for tin, zinc, and antimony as well.[126] Recent studies in freshwater environments have shown that lead can also be biologically methylated.[127,128] Evidence has been presented more recently that particles containing high concentrations of zinc can be released into the atmosphere by growing plants.[90]

It has been estimated that there is a total mass flux of crustal materials to the atmosphere of 2.5×10^{14} g/year,[129] although recent estimates suggest this value may be too low by at least a factor of 6. The flux of sea salt particles to the atmosphere has been estimated by Eriksson[130] at 10^{15} g/year, with Blanchard's estimate[69] higher by a factor of 10. Using the two values given above, the average crustal concentration given by Taylor,[131] and the average seawater concentrations of metals reported by Riley[132] and by Chester and Stoner,[133] an estimate of the metal transferred to the atmosphere was calculated by the Working Group on Metals at the Miami Workshop.[9] It has been assumed that no elemental fractionation occurs during the sea-salt particle formation. The contribution of heavy metals from the combustion of coal, lignite, oil and natural gas to the atmosphere has been evaluated by Bertine and Goldberg.[134] Patterson et al.[135] estimated the flux of submicrometer lead to the atmosphere from anthropogenic sources, largely the combustion of leaded gasoline. The estimated global fluxes of metals to the atmosphere from all these sources is given in Table 4.

Table 4
GLOBAL FLUX OF METALS TO THE ATMOSPHERE BASED ON TOTAL CRUSTAL MATERIAL FLUX OF 2.5×10^{14} G YR^{-1} [a] AND TOTAL SEA-SALT FLUX OF 1×10^{15} G YR^{-1} [b]

Element	Crustal material[c] 10^9g yr^{-1}	Bulk sea salt[d] 10^9g yr^{-1}	Fossil-Fuel combustion products[e] 10^9g yr^{-1}
Al	20,000	0.15	1400
Fe	14,000	0.5	1400
Na	6,000	3×10^5	300
Mn	200	0.005	7
Sc	6	0.000015	0.7
Cu	14	0.04	2
V	30	0.05	12
Se	0.013	0.003	0.05
Pb	3	0.0008	150[f]
Cd	0.05	0.0008	—
As	0.5	0.05	0.7
Zn	18	0.08	0.5
Sb	0.05	0.007	—
Hg	0.02	0.0005	1.6

[a]   Goldberg.[129]
[b]   Eriksson.[130]
[c]   Using crustal abundances of Taylor.[131]
[d]   Using seawater concentrations of Riley[132] and Chester and Stoner.[133]
[e]   From estimates of Bertine and Goldberg.[134]
[f]   Estimate of Patterson et al.[135]

From NAS, *The Tropospheric Transport of Pollutants and Other Substances to the Oceans,* Workshop on Tropospheric Transport of Pollutants to the Ocean Steering Committee, Ocean Sciences Board, Assembly of Mathematical and Physical Sciences, National Research Council, National Academy of Sciences, Washington, D.C., 1978.

Obviously, with the level of uncertainty of the source strength estimates themselves, the crustal material and bulk sea-salt values are only order-of-magnitude estimates. If significant chemical fractionation of these metals occurs during bubble bursting, the fluxes calculated from bulk sea-salt concentrations will be too low. In fact, the estimate of metal ejection into the atmosphere from anthropogenic sources is probably considerably more accurate than that from crustal weathering, the ocean, or any other natural sources.

There are now numerous stations covering the globe in the urban terrestrial environment at which samples are collected for metal analysis. Unfortunately, the sampling of the atmosphere above the oceans is not a simple matter, and data from this source are still rather scanty. However, a few island stations have been occupied over the past decade, and measurements of metal concentrations are becoming available for samples taken at these stations. These include Bermuda and the Shetland Islands in the Atlantic, and Oahu of the Hawaiian Chain in the Pacific. Additional samples have been or are being collected on Enewetak Atoll and American Somoa of the Equatorial Pacific. However, most of the recently available data on the concentrations of metals in the marine atmosphere are for the north Atlantic. There is very little information for the atmosphere of the oceans in the southern hemisphere, although recent studies are beginning to fill this gap.[93] Some data are available for the south pole.[136,137] Concentrations of metals in the marine atmosphere and in the sea water at a number of these stations are given in Table 5.

The flux of metals from the atmosphere to the global ocean cannot be estimated accurately at present because of a complete lack of data from many areas, particularly in the southern hemisphere. Even in marine areas for which data on atmospheric concentration of metals are available, no direct flux measurements for metals have been made. An estimate of the air-to-sea flux of many metals in the North Atlantic has been made by the Working Group on Metals at the Miami Workshop.[9] A mean atmospheric concentration for each metal over the North Atlantic was estimated from concentration measurements at Bermuda and over the eastern tropical North Atlantic. The model assumes that the particles containing the metals are distributed uniformly from the sea surface to 4,000 m altitude and that the particle population in this part of the atmosphere is removed 50 times a year. The calculations yield only a rough approximation of the metal flux in that area.

Metals of potential concern because of their pollution significance have been identified as lead, mercury, arsenic, selenium, antimony, zinc, cadmium, copper, nickel, chromium, and vanadium.[138] Some of these elements have been selected on the basis that the atmospheric concentrations exceed those derived from the ratio of natural concentrations to concentrations predicted from crustal weathering or bulk seawater sources. Others have been identified as environmentally hazardous mainly in occupational exposure. All these elements and their compounds may be involved in air-sea interchange. Mercury, cadmium, and lead have received the most attention because of their human health effects. Iron and aluminum have been designated as crustal composition references, and sodium as a seawater composition reference element, contributing to the determination of atmospheric transport.[143] The enrichment factor for a metal X in an aerosol, using sodium as the reference element, is defined as

$$\mathrm{EF}_{sea} = \left(\frac{X}{Na}\right)_{air} \bigg/ \left(\frac{X}{Na}\right)_{sea}$$

where the numerator and denominator are, respectively, the ratios of the concentration of the metal X to that of sodium in the atmosphere and in bulk seawater.

D. Radionuclides

The atmospheric transfer of certain radionuclides, stemming from nuclear weapons tests, has been studied intensively.[4] World-wide monitoring for fallout was carried out for many years following the early atmospheric and sea-surface weapons tests and continues to the present on a much reduced scale. Such fission-product nuclides as ^{137}Cs and ^{90}Sr have been followed also in the marine environment for the purpose of tracing water movements and determining the rate of exchange.[144] With the 1963 Test Ban Treaty, the major nuclear powers have ceased atmospheric and oceanic testing of nuclear weapons so that input of radionuclides into the atmosphere has steadily declined since then. However, newer nuclear powers, which are not party to the Test Ban Treaty, have begun testing nuclear weapons, and there have been brief episodes of increases in radioactive fallout recently.

The environmental radioactive problem today, however, revolves mainly around peaceful uses of radioactivity. In normal operation, nuclear power reactors are considered to be comparatively radiation free. Although the possibility of nuclear accidents has been virtually ruled out by nuclear power authorities, the several recorded reactor accidents, particularly the one during the 1950s involving atmospheric emissions of iodine-131 at Windscale in the U.K. and the more recent accident at the Three Mile Island nuclear facility near Harrisburg, Pennsylvania, continue to be stark reminders for the public that accidents can happen.

Nevertheless, the issue of greatest concern to the public is reprocessing of nuclear wastes and the possible loss by accident, theft, or terrorist activities of plutonium arising from such reprocessing. Plutonium, a major ingredient for manufacture of nuclear weapons, is not only highly toxic through its radioactivity, but is considered to be one of the most potent carcinogens known to man. The greatest experience on behaviour of plutonium-238 in the atmosphere was gained from the investigation of the fallout from the accidental combustion due to rocket failure after launch of a navigational satellite with a nuclear power source (SNAP [Systems for Nuclear Auxiliary Power] — 9A package) on April 21, 1964.[145] The disintegration of the SNAP device doubled the amount of plutonium-238 which has been added to the environment by nuclear weapons testing.

It is estimated that by the year 1990 there will be about 50 tons of plutonium in the transport mode in the U.S., and perhaps double or triple that amount globally.[139] There will also be a substantial amount of americium and curium in the fuel cycle of nuclear reactors. The question that might be posed is what the leakage will be to the environment and whether this will be mainly into the atmosphere or into aquatic systems. The human health implications of nuclear power production have been well summarized in a recent World Health Organization report.[146]

Besides the radionuclides noted in the foregoing, there is expected to be continued input into the environment of such radionuclides as carbon-14 and tritium, although there is no reason to expect that these will pose ecological hazards at present rates of introduction. It should be noted that there are naturally occurring nuclides which pose a natural or background burden of radiation to organisms, the most important of which are potassium-40 and rubidium-87. With respect to natural background dose rate to all groups of organisms, the alpha-emitting nuclides, particularly polonium-210, are the most significant. The distribution of natural radioactivity in association with atmospheric aerosol particles has been recently assessed.[147]

The net flux of radionuclides in open ocean areas is generally considered downward from the atmosphere into the sea. It is conceivable that there is a small flux from the sea to the atmosphere mediated by biological processes. *Sargassum,* the seaweed common to the Sargasso Sea, appears to be an efficient collector of plutonium.[139] This seaweed-adsorbed plutonium can be redistributed by the drift of these plants and might even be concentrated in coastal waters. On decay of the seaweed, the plutonium may be released into the surface water and injected into the atmosphere by various sea-to-air transfer mechanisms. High levels of plutonium and polonium have been found on the surface of other large marine algae, e.g., the giant brown eel kelp, *Pelagophycus porra,*[148] and the giant brown kelp, *Macrocystis pyrifera.*[149]

E. Gases

1. Carbon Dioxide

On a global basis, the major gaseous contribution of man is carbon dioxide. There has been a finite increase in concentration of atmospheric carbon dioxide associated with the combustion of fossil fuels[150-152] from less than 300 ppm in the pre-industrial era to over 330 ppm in 1976. While the increased carbon dioxide is not expected to have direct adverse effect on human health or on other life on this planet, it is generally agreed that it could have an impact on climate and associated phenomena through increasing temperatures from the so-called "greenhouse effect". However, climatologists are not yet agreed on the magnitude of the carbon dioxide effect in increasing

Table 5

POLLUTANTS INVOLVED IN EXCHANGE BETWEEN THE ATMOSPHERE AND THE SEA

Pollutant category	Pollutant	Enrichment factor[a]	Atmosphere (location)	Ocean (location)	Ref.
				Conc range	
Halogenated hydrocarbons	PCB		0.2—0.6 ng m⁻³ (Bermuda: 30°20'N; 64°40'W)	0.4—1.9 ng L⁻¹ (Atlantic: 30°25'N; 70°20'W)	138
	DDT		0.02—0.06 ng m⁻³ (Bermuda: 30°20'N; 64°40'W)	0.4—0.5 ng L⁻¹ (Sargasso Sea: 29°56'N; 64°40'W)	139
					138
					2
	Chlordane		0.01 ng m⁻³ (Bermuda: 30°20'N; 64°40'W)		138
Petroleum hydrocarbons	boiling range, n-C₁₄ to n-C₃₂		60—600 ng m⁻³ (Bermuda; 30°20'N; 64°40'W)	0.5—6 µg L⁻¹ as "hydrocarbons" (near Bermuda: 32°18'N; 65°32'W)	139
Metals			Conc in air as atmospheric particulates Bermuda		
	Aluminum	1.0	130 ng m⁻³	10 µg L⁻¹	17, 140, 142
	Manganese	1.5	1.5 ng m⁻³	2 µg L⁻¹	140
	Iron	1.1	100 ng m⁻³	10 µg L⁻¹	142
	Cobalt	1.8		0.05 µg L⁻¹	141
	Vanadium	17	0.16 ± 0.11 ng m⁻³	2 µg L⁻¹	140
	Zinc	20		2 µg L⁻¹	141
	Copper	10	1.9 ± 2.1 ng m⁻³	3 µg L⁻¹	140
	Cadmium	500		0.11 µg L⁻¹	141
	Antimony	180		0.45 µg L⁻¹	112
	Lead	180	3.0 ± 2.8 ng m⁻³	0.03 µg L⁻¹	140
	Selenium	2600		0.45 µg L⁻¹	112
	Arsenic	30	0.08 ng m⁻³	2.6 µg L⁻¹	141
Radionuclides				North Atlantic Ocean	
	Strontium-90			0.13 (0.02—0.50)[b] pCi L⁻¹	139
	Cesium-137			0.21 (0.03—0.8) pCi L⁻¹	
	Tritium			48 (31—74) pCi L⁻¹	
	Carbon-14			0.02 (0.01—0.04) pCi L⁻¹	
	Plutonium-239			(0.3—1.2)×10⁻³ pCi L⁻¹	

Microorganisms	Bacteria	Seawater bacteria enriched 10—100 times in droplets from bursting bubbles over conc in bulk seawater (laboratory); viruses from surf enriched 200 times in aerosol	0—10^6 fecal coliforms (100 mL)$^{-1}$ depending on proximity of raw sewage or sludge input	18, 94
	Viruses			20
	Fungi			

[a] Enrichment Factor (EF) is calculated for any element (X) in the atmosphere relative to the crust using aluminum (Al), which is present in high concentrations in the crust, as a reference element.

$$EF = (X/Al)_{atm} / (X/Al)_{crust}$$

[b] Average values with range given in brackets.

atmospheric temperature, and how much of this will be cancelled by the effect of particulate matter in the global atmosphere, which decreases the amount of solar radiation reaching the earth and thereby depresses the temperature of the earth. Depending on the model used in computations, a doubling of the atmospheric CO_2 concentration can increase the surface temperature of the earth by 1.5 to 3°C, which is twice the magnitude of temperature fluctuations over the last century.[10,153]

The sea probably plays a role in moderating the effect of increased carbon dioxide through air-sea exchange of this gas. Exactly what the effect, if any, of the increased carbon dioxide is on the oceans is still debated. At least in shallow seas, the carbon dioxide-bicarbonate-carbonate balance could be affected. Plant life in the sea as on land utilizes carbon dioxide in photosynthesis and releases it during respiration. Marine plants and animals release CO_2 during decay.

The air-sea transfer of CO_2 is controlled by processes in the liquid phase,[154] and it is possible in principle to calculate the transfer of CO_2 through the air-sea interface knowing the state of saturation of the surface water with respect to atmospheric CO_2 levels. Unfortunately, there is a great deal of spatial variability in the CO_2 saturation of surface waters and many data would be required to evaluate the net CO_2 flux. A suggested more effective way of determining the CO_2 uptake by the oceans is by utilizing the distribution of natural and nuclear-weapons-produced ^{14}C in the atmosphere-ocean system.[139]

There are basically 4 pools of carbon on our planet: (1) atmosphere, 70×10^{10} T as carbon dioxide; (2) world-wide biota, 80×10^{10} T; (3) organic matter of the soil (humus and peat), 100×10^{10} to 300×10^{10} T; and (4) oceans, 4000×10^{10} T. All these pools have an interchange of carbon, but the rate of exchange between the atmosphere and the oceans as a whole is considered to be low. The most rapid exchange occurs between the atmosphere and the upper 100-m mixed layer of the sea, where there is a reservoir of 60×10^{10} T of inorganic C. Currently, there is an annual injection into the atmosphere of CO_2 from fossil fuels amounting to 0.5×10^{10} T of C, which leads to an annual atmospheric increase of 0.23×10^{10} T of carbon. This leaves 0.27×10^{10} T of fossil-fuel C to be removed by some unknown combination of terrestrial and oceanic processes.[155] The net effect of oceanic processes in buffering the input of fossil-fuel CO_2 in the atmosphere has not yet been established.

2. Sulfur Dioxide

The presence of sulfur dioxide in the atmosphere from smelters, fossil fuel combustion, emissions from other industries, and from natural sources can seriously affect freshwater ecosystems and particularly the fish life, as already identified in the Scandinavian countries.[11-15,156-168] The problem of acidic precipitation due to sulfur dioxide is unlikely to affect the marine environment because of the high buffer capacity of sea water. However, it is conceivable that acid precipitation could locally and temporarily affect the brackish seas, such as the lower salinity areas of the Baltic, and surface layers of certain fjords and estuaries, where there is a large inflow of fresh water at the surface, and salinity remains low in the nearshore waters.

Using a value of 3 μg m^{-3} for the background concentration of SO_2 in marine air, Liss and Slater[53] calculated a total flux of SO_2 from the atmosphere to the oceans as 1.5×10^{14} g yr^{-1}. More recently, however, Meszaros[105] concluded from a review of available measurements that the background SO_2 level should be approximately 0.1 to 0.2 μg m^{-3}. In a reevaluation,[102] a value lower by an order of magnitude from the original is given for the global flux of SO_2 to the oceans at approximately 10^{13} g yr^{-1} (Table 3).

Biogenic sulfur-containing compounds are released into the atmosphere both from land-based sources and the sea.[169-172] However, no major natural sources of SO_2 have been identified.

3. Halogens

Because of their release into the atmosphere and into various aquatic environments from numerous industrial processes and because of their toxicity, halogens are considered to be of some importance in air-sea exchange processes.

a. Chlorine

Chlorine is well known as a highly toxic gaseous element which is transported in large quantities by land and sea, and can be extremely hazardous to human health in the event of accidental release. It has been identified for consideration in international forums owing to its atmospheric pollution significance.[143] Chlorine is used in the pulp and paper industry for bleaching, in sewage and drinking water for disinfection, as a biofouling preventive agent in seawater cooling lines, and in a host of other applications requiring smaller volumes. There are chronic releases of chlorine into the environment both deliberately for disinfection and biofouling control, and as a result of leaks from manufacture and industrial application. The risks of accidental release are always present, and many incidents of evacuation of people following chlorine tank car derailments can be cited. There have also been instances of chlorine-transporting vessels, posing a threat of emission of poisonous gas in the event of collision or grounding, and tank cars of chlorine have been lost from barges at sea. Because chlorine is heavier than air, it lies close to the ground or sea surface when released and presents its greatest hazard to air-breathing animals early following release. Residual chlorine is usually present in sewage effluent if disinfection by chlorination is practiced, and in drinking water and swimming pools, and some of this is undoubtedly released to the atmosphere.

Being a highly reactive gas, chlorine does not remain long in the nascent form. It is rapidly reduced in the environment. Hence the flux of this gas between the atmosphere and the sea could only be evaluated for rather transient conditions in local areas. Observations have been made on particulate chlorine distribution in the atmosphere of an industrialized part of Texas.[173]

b. Bromine

Bromine is used industrially for gold extraction, bleaching of silk and other fibers, and for the manufacture of medicinal bromine compounds, dyestuffs, and anti-knock compounds for gasoline motors. It has been proposed as a more powerful oxidizing agent than chlorine in certain disinfectant and industrial bleaching applications. Like chlorine, it is highly reactive and does not remain long in the elemental form in the environment. Bromine has also been identified as having atmospheric pollution significance.[143]

c. Fluorine

Fluorine is an important component of emissions from electrolytic reduction in the production of aluminum and phosphorus. Because it poses a health hazard to humans as well as to domestic animals and has been shown to be detrimental to vegetation, including forests, it should be considered as at least a hazardous atmospheric pollutant. No evidence has been presented, however, that atmospheric fluorine emissions pose a hazard in marine waters, although sampling and analysis of waters in a marine inlet receiving discharges from an aluminum smelter showed elevated levels of fluoride.[174]

4. Other Gases

Global fluxes of a number of gases — N_2O, CO, CH_4, CH_3I, $(CH_3)_2S$ — whose net transfer is from the oceans to the atmosphere, have been calculated by Liss and Slater.[53] Undoubtedly, biological processes are involved in the production of these gases.

Their estimated fluxes are given in Table 3. The concentration of CH_4 over the ocean and its possible variation with latitude has been considered recently.[175] Vertical profiles of a number of substances (H_2, CH_4, CO, N_2O, $CFCl_3$, and CF_2Cl_2) have been reported for the midlatitude stratosphere and troposphere.[176]

F. Microorganisms

The exchange of microorganisms between the atmosphere and the sea is seldom considered in terms of pollutant transfer. The injection of microorganisms from the atmosphere to the sea would be inconsequential when compared with the input from ocean dumping and sewer outfalls. The potential transfer of pathogenic bacteria and viruses from the sea to the atmosphere, however, does merit examination.[94]

It has been demonstrated conclusively that viruses can be enriched in the surface microlayer by bubbles and projected into the atmosphere by bubble bursting and from surf to wind.[18-20,76,78,80] At the present state of knowledge, it is uncertain how epidemiologically significant the transfer of pathogenic microorganisms from the sea to the land through the atmosphere might be, although convincing evidence has been presented for the transmission of *Mycobacterium intracellulare* (BATTEY) infection from coastal Atlantic waters into the southeastern continental U.S.[21] In most cases, an immunity to disease microorganisms is built up through continuing exposure. It has been suggested that the danger to health of people living along coasts from airborne transmission of pathogens lies in an epidemic situation where no previous immunity has been developed.[20]

In addition to transport of bacteria and viruses from sea to land via the atmosphere, there is potential for transmission by this mode of spores of fungus and parasites that could infect not only humans and other land animals, but also agricultural produce. The greatest threat in transmittal from the sea to the atmosphere of disease microbes lies in those microorganisms which are resistant, or are rendered so by oil films, to desiccation and UV light, or can form resistant spores or both. However, little is known about transmittal of disease organisms through the atmosphere, except possibly in the classified literature dealing with research on biological warfare.

It has been found in laboratory experiments with both freshwater and seawater bacteria, that high surface bacterial concentrations are reflected in the drops ejected from bursting bubbles.[76,78,177,178] Concentration factors of 10 to 100 were noted for the bacteria, and in sea water they were considerably higher for natural bacterial populations than for a pure culture of *Serratia marinoruba.* When bubbles rose 1 cm through a bacterial suspension of the freshwater species *S. marcescens,* the concentrations in the jet drops were 10 to 100 times that in the bulk water. A bubble ascent of 100 cm or more led to concentration factors in excess of 1000.[76] Utilizing Hudson River water, Blanchard and Syzdek found that for bubbles rising through 1 and 4 cm of water, the concentration factors for several species of bacteria in the airborne drops were about 10 and 35, respectively.[76]

In studies of virus transfer from the sea surf to the atmosphere, it was found that bubble levitation of viruses injected into the surf produced 200 times more viruses mL^{-1} in the aerosol than were present in samples from the surf.[20] The increase of viruses in the aerosol over that in the bulk seawater is caused by the adsorption of viruses to air bubbles as they rise through the water, and the stripping of the virus-rich bubble surface and ejecting it into the atmosphere when the bubble bursts. The frequency of virus-bearing drops decreased exponentially with distance downwind from the surf.

It has been estimated for the New York area that if the surf alone produces 3×10^5 bubbles m^2 sec^{-1},[74] and if it is assumed that the surf has an area of 2.5×10^6 m^2 (25 m wide by 100 km long) there will be 7.5×10^{11} bubbles generated per second. Assuming that each bubble ejects a single bacterium, there will be 7.5×10^{11} bacteria lofted into the atmosphere per second. With an introduction of 5×10^{14} bacteria sec^{-1} by the Hudson River, and if this water were evenly distributed along the 100 km of surf, it would require less than 12 min to inject the bacteria into the atmosphere at the above rate of removal.[20]

IV. POTENTIAL PROBLEMS DUE TO INTERCHANGE OF POLLUTANTS BETWEEN THE ATMOSPHERE AND THE SEA AND RESEARCH NEEDS

A. Halogenated Hydrocarbons

The ubiquitous nature of DDT and the PCBs and their presence in marine organisms in virtually every corner of the world oceans have been attributed, with sound scientific support, to transport by the atmosphere. These chlorinated hydrocarbons, along with many others, are used in a variety of ways by man, often leading to their release deliberately or inadvertently into the environment with the oceans as the sink. Because they are not normally present in nature, ecological disruptions can occur when they are artificially introduced. Reproductive failure due to eggshell thinning in predatory birds has been well documented for DDT and its metabolites. This pesticide is not only toxic to young fish but also causes reproductive failure in certain fish species. Similar effects of the PCBs have been observed, and there is evidence of acute toxicity to some aquatic organisms in extreme exposure situations. A concern sometimes expressed is that synthetic compounds, alien to the marine environment, may inhibit certain vital biological processes in the sea. These processes might be associated with reproduction of a given species, and there is always a danger of widespread reproductive failure in such species and an ultimate decline in populations. It has even been suggested[139] that the lower molecular weight halogenated hydrocarbons, such as the chlorofluorocarbons, may inhibit the normal biochemical cycles involving bacteria, and thereby retard degradation of organic matter.

B. Petroleum Hydrocarbons

Petroleum hydrocarbons do not pose the same ecological threat to the marine environment as the halogenated hydrocarbons. After all, there have been natural seeps of petroleum hydrocarbons for millions of years, and no ecological disasters have yet been identified. Nevertheless, the long-term ecological impact of an oil spill, especially with different chemical characteristics of the spilled oil and different environmental conditions, is still not fully known. The flux of petroleum hydrocarbons to the oceans via the atmosphere probably still makes only a small contribution to the marine burden of these substances compared to direct input from tankers, shore installations, and industrial and domestic sewer outfalls. The presence of oil films, however, could exacerbate other pollution problems, such as providing a medium for concentration of chlorinated hydrocarbons and metals.

C. Metals

Heavy metals, similar to the petroleum hydrocarbons, are not entirely alien to the marine environment. Although some are present in only minute concentrations, virtually every metal can be detected in sea water. The anthropogenic input of metals into the marine environment merely leads to higher seawater concentrations. It was thought at one time that man plays a major role in the mercury budget of the oceans, but this is now questionable.[140]

Several metals have been identified, however, where anthropogenic input apparently approaches within an order of magnitude, e.g., mercury and barium,[134] or exceeds that arising from the natural geochemical cycle, e.g., lead.[135] From among the metals added to the oceans in appreciable quantities by man's activities, lead probably merits the greatest concern. Its presence in greater than background levels can be identified in surface marine waters adjacent to major population centers, e.g., the north Pacific Ocean along the southern California coast[179] and off the U.S. Atlantic coast,[180] where undoubtedly the atmosphere has played an important role in transporting the lead from automobile emissions and other sources on land to the sea. The increased use of coal to supplement dwindling petroleum and gas supplies for energy will probably result in large inputs of metals into the atmosphere. It is difficult to predict on the basis of present knowledge, however, whether this will lead to significant increases in the metals burden of the marine environment.

D. Radionuclides

In the absence of resumption of atmospheric testing of nuclear weapons and/or a catastrophic input of radioactivity from nuclear accidents or a nuclear war, radioactivity in the atmosphere and the marine environment can be expected to stabilize at a comparatively low level. Nevertheless, research is justified on the transfer through the environment of certain hazardous, alpha-emitting radionuclides such as plutonium, polonium, and other transuranics. The accidental release of these nuclides in peaceful uses of atomic energy could pose special ecological and environmental health problems. A better understanding of their behavior in the environment and the mode of transfer from one component of the global ecosystem to another would allow better management of any exigency involving such nuclides.

E. Gases

The incineration of wastes at sea poses a problem of unknown magnitude. With plans of some countries to burn such unwanted materials as PCBs and other organohalogens on incinerator ships, there will be emission of waste products of combustion into the atmosphere and ultimately entry by rainout and washout into the sea. Again, local effects could arise, although controls are being introduced to minimize any environmental damage.

The "tall stack" policy in many countries to remove atmospheric emissions from the immediate vicinity of industrial plants and cities (e.g., Sudbury, Ontario) has led to increased injections of pollutants into the upper atmosphere, where they can be rapidly transported and dispersed by strong high-altitude winds. While offending emissions so far have been identified mainly as those affecting terrestrial resources and fresh waters, i.e., sulfur dioxide, sulfates, and oxides of nitrogen, substances damaging to the marine environment could conceivably also be released through such high stacks.

The injection of pollutants into the sea from the atmosphere in the open ocean may have a comparatively minor effect, because of the large volume of water involved; but in confined inshore waters, the impact could be severe. For example, the atmospheric emissions from a large plant at the head of a fjord could conceivably have a marked effect on water characteristics at its head, if the prevailing winds are in that direction. Because the surface water of a fjord is comparatively fresh and low in salinity, its buffering capacity may be quite low. Therefore, atmospheric input of sulfur dioxide could lead to a lowering of pH and possible harm to the living resources. The input of atmospheric phosphate and oxides of nitrogen could add an undesirable nutrient load to a coastal marine system. Thus, local impact could be far more severe than any detectable global effect. Although the global trends are the most important ones from the long-term point of view, the effects in fresh waters and in nearshore coastal waters can provide an early warning system for oceanic impact. Once trends in oceanic waters can be recognized, it may take a long time to reverse them.

F. Microorganisms

It was noted in the report "Disposal in the Marine Environment",[18] that ". . . it is known that sewage treatment plants, cooling towers, and tanneries, for example, produce bacterial aerosols that travel into outlying communities and have produced viral disease. Many of these aerosol particles produced by bubbling and splashing of liquids are in the size range (1 to 5 μm) where maximal alveolar penetration occurs". The water-to-air transfer of pathogens is of considerable local concern in some areas receiving raw sewage and sewage sludge.

G. Processes

There is a dearth of information on processes involving pollutants both in the atmosphere and in the sea. However, the greatest lack of information probably concerns the air-sea interface itself and the various exchange processes that occur through the surface microlayer. Such scientific gatherings as the 1975 Miami Workshop on Tropospheric Transport of Pollutants to the Ocean[9] help to bring together available information and identify the gaps in knowledge. The SEAREX program[30] will undoubtedly fill many of the gaps that now exist in knowledge on air-sea exchange of pollutants.

The bursting bubble has been shown to be the major process for transferring seawater and inorganic constituents contained in it into the atmosphere. It has been suggested that the bubble mechanism is also responsible for water-to-air transfer of organic material, particularly the surface-active components, and metals. This can have certain implications for coastal vegetation, although the seriousness on a global scale is not fully appreciated.

There are still many unknowns in the field of air-sea exchange of pollutants, particularly in connection with processes in the atmosphere, the sea, and the air-sea interface. Some of the research needs to fill the apparent gaps in knowledge are listed in Table 6.

V. CONCLUSIONS

The long-range transport of certain pollutants through the atmosphere is a serious problem in certain sensitive freshwater environments. At the present time, the principal harm identified is related to sulfur dioxide and its effects on freshwater and forest ecosystems in the Scandinavian countries, northeastern U.S., and in eastern Canada. The high buffering capacity of seawater precludes the large changes in pH that occur in soft freshwaters from acidic and alkaline wastes.

No acute conditions resulting from atmospheric input of pollutants, similar to those found in freshwaters, have been identified in the marine environment. However, the world-wide distribution of DDT and PCBs in the marine environment, as recognized from the presence of these chlorinated hydrocarbons in fatty tissues of marine organisms from every quarter of the globe, has been attributed in part at least to atmospheric input. The problem of petroleum hydrocarbons in the sea, arising from direct spills into the marine environment, may be exacerbated by atmospheric input. The net flux from oil spills is generally upward with evaporation of the lighter fractions of the oil. The increased atmospheric burden of carbon dioxide from combustion of fossil fuels may be reflected by changes in the oceans, but the effect is probably one where the air-sea exchange ameliorates the atmospheric levels of this gas. The net pollutant CO_2 flux is downward in the open ocean, but the direction of net flux of natural and pollutant CO_2 combined is unknown.

Input of certain metals from the continents has been accelerated by man's activities, but anthropogenic lead is the only one that appears to be causing concern at the present time. Radionuclides from anthropogenic sources have been measured in the atmos-

Table 6
RESEARCH NEEDS ON DIFFERENT ASPECTS OF INTERCHANGE OF POLLUTANTS BETWEEN THE ATMOSPHERE AND THE SEA

Environment	Research Needs
Atmosphere	Measurement of fluxes of different pollutants through the atmosphere
	Methods to distinguish fluxes of natural and anthropogenic contributions of materials to the atmospheric burden
	Distribution of pollutants between vapor phase and particles in the marine atmosphere
	Behavior in the atmosphere and marine environment of low molecular weight hydrocarbons, hexachlorobenzene, and heterocyclic petroleum compounds
	Elucidation of chemical forms of substances in the atmosphere and suitable techniques for sampling them
	Transformation of substances in the atmosphere by photochemical oxidation, etc.
	Mechanism of removal of pollutants, e.g., Freons and other halocarbons, from the atmosphere and deposition into the sea
	Amounts of such pollutants as metals and halogenated hydrocarbons present in precipitation and dry fallout in open ocean areas
	Rate of fallout (wet and dry) of heavy fractions of hydrocarbons
	State of aggregation of volatile metals in air
	Models on movement and behavior of substances in the atmosphere
Oceans	Determination of chemical speciation and stability of different pollutants in the marine environment
	Techniques for routine differentiation between biogenic and petroleum hydrocarbons
	Measurement of fluxes of pollutants through different pathways in the marine environment
	Effects of point sources of pollution on air-sea exchange
	Marine ecological impact of low molecular weight hydrocarbons and chlorinated hydrocarbons, hexachlorobenzene, and heterlocyclic petroleum components
	Bioaccumulation of pollutants by marine organisms and biomagnification through the food chain
	Association of trace metals with organic substances and/or mineral matter of particles in the sea
	Physical transfer of metals in the particulate and dissolved phases and estimates of residence time
	Ultimate fate of certain critical pollutants
	Formulation of predictive models on transfer of pollutants between different reservoirs in the marine environment
	The effect of marine biota, especially in upwelling areas, on release of metals and organic substances into the atmosphere
Air-sea interface	Sampling and analysis for critical pollutants in the surface microlayer of areas having different characteristics
	Rate of reemission to the atmosphere from the sea surface of certain critical pollutants
	Transfer of bacteria and viruses from the sea surface to the atmosphere
	Measurements of bubble size and number distribution as a function of wind speed in the open ocean
	Investigation of the surface microlayer as an ecosystem for marine microflora and microfauna
	The characteristics of the air-sea interface which enhance concentration of certain pollutants in the surface microlayer and inhibit gaseous exchange
	Development of the theory of transport across the air-sea interface, particularly under conditions of turbulence and rapid bubble formation and breaking
	The physical effects of pollutant organic films on properties of the air-sea interface and on exchange processes
	The effect of micoorganisms in the microlayer on air-sea exchange of metals and organic constituents

phere and the sea for a long time, partly because of public concern for these radioactive materials and partly because sensitive techniques for their measurement have been available. Plutonium and other alpha-emitting transuranics appear to be causing the greatest public and scientific worry at present. More research on their behavior in the sea would at least help to allay fears about these dangerous nuclides.

While the impact on freshwater ecosystems of certain airborne pollutants may have little bearing on the marine environment, the atmospheric transport and air-water exchange processes are similar. Therefore, it would be profitable for marine workers on air-sea exchange of pollutants to exchange information and perhaps even collaborate with freshwater investigators in this field. There are certain problems unique to the marine environment, however, and many gaps in information on processes of air-sea exchange of pollutants still exist. Models of transfer of pollutants in both the atmosphere and the marine environment are needed to provide a predictive capability on their dispersal and fate.

The exchange of pollutants at the air-sea interface, particularly the transfer from the water to the atmosphere, requires some detailed investigation. With improved techniques available for sampling the surface microlayer, a program of sampling of areas having different characteristics and pollutant inputs should be conducted. The effects of pollutant surface films on properties of the air-sea interface and the exchange processes should be further studied. Processes in the sea which concentrate certain pollutants near the surface (particularly bacteria and viruses in certain sewage and sludge dumping areas) and make them available for entry into the atmosphere should be investigated more fully. Bubble formation and bursting at the air-sea interface are considered to be important contributors to concentration of pollutants and ejection from the sea to the atmosphere. Further studies on the hydrodynamics of bubble formation are needed not only for basic knowledge but also for effective management of marine waste disposal and pollution control. The impacts of rain and snow on the surface microlayer and on upward transfer of pollutants from the sea need further study, particularly to compare the net effect on such transfer with that of bubble bursting. The effects of marine biota on air-sea exchange of pollutants is virtually unknown and merit detailed investigation.

ACKNOWLEDGMENTS

This review was initially prepared as a background paper for the first session of the GESAMP (IMCO/FAO/Unesco/WMO/WHO/IAEA/UN Joint Group of Experts on the Scientific Aspects of Marine Pollution) Working Group on Interchange of Pollutants between the Atmosphere and the Oceans held at Dubrovnik, Yugoslavia, October 3 to 7, 1977. I thank members of the Working Group for their assistance on the first draft, individually and in discussion at the first and second sessions, and experts at the tenth session of GESAMP, Unesco, Paris, May 29 to June 2, 1978, for their comments on the draft reviewed in that session. Mrs. Donna Price deserves honorable mention for her perseverance in retyping the manuscript many times.

GLOSSARY

Absorption — The process of entry usually by a dissolved substance into the inner structure of another substance or organism. This may be either physicochemical, as in the case of a liquid taking up molecules of a gas or vapor, or physiological, as in the passage of nutrients into the bloodstream through the intestinal walls or other tissue.

Adsorption — The process of adhesion of gaseous, dissolved or particulate constituents to the outer surface of a living or inanimate object, sometimes called the adsorbent.

Advection — Movement of water and substances contained therein by currents vertically or horizontally.

Aerosol — Particles, dry or wet, in the atmosphere which can be collected on an appropriate filter.

Alveoli — The tiny sac-like cavities lining the inside of the lungs and other parts of the respiratory system.

Ambient — Pertaining to surrounding space in the atmosphere or the sea.

Anthropogenic — Man-made or resulting from the activities of man.

Biofouling — The encrustation of objects immersed in or exposed to water by aquatic organisms, e.g., barnacles, mussels, and algae in the marine environment.

Biogenic — Pertaining to or produced by plant or animal organisms.

Brownian movement — Random motion of tiny particles, suspended in air, water or other fluid medium, generated by molecular collisions.

Buffering capacity — The ability to resist large, abrupt changes in hydrogen-ion concentration, represented by pH, provided by weak-acid salts, e.g., bicarbonates and borates in seawater. May relate also to ability to resist or ameliorate other changes, e.g., CO_2 concentrations in the atmosphere.

Burden — The total load of a substance in all or part of the atmosphere or the sea.

Capillary waves — Tiny gravity waves on the water surface, ranging from a barely visible size to a few millimeters in height, usually generated by a very light breeze.

Chlorinated hydrocarbons — Organic substance with chlorine atoms attached to specific locations on the molecule.

Coagulation — The physicochemical process of clumping or aggregation of particles to form clusters or flocs in both the atmosphere and the sea.

Convection — Motion generated by uneven heating or other density-disturbing processes in the atmosphere or the sea.

Crustal weathering — The removal of components of the earth's crust by atmospheric processes, e.g., precipitation, runoff, freezing, and evaporation.

Dedicated computer — An electronic data processing unit assigned to a specific purpose, e.g., analytical data processing.

Deposition velocity — The rate with which particles are deposited on the land or water surface per unit area.

Desiccation — The process of drying out.

Diffusion (molecular) — The random motion of matter at the molecular scale with a net transfer of material in the direction of decreasing concentration.

Diffusion (eddy) — The motion of matter on a larger scale related to the size of eddies. This is the usual form of diffusion experienced in the atmosphere and the sea.

Dispersion — A collective term for the spread of material in the atmosphere or the sea by various physical processes, e.g., winds, currents, turbulence.

Emulsification — The suspension of one immiscible substance in another, e.g., the result of mixing oil into water.

Enrichment factor — The concentration of a substance in one medium in relation to that in another, usually expressed as a ratio, e.g., the ratio of the ratio of copper to aluminum in the atmosphere to that in the earth's crust.

Fallout — The gravitational deposition of particulate matter. Wet particles contribute *wet fallout,* while dry particles lead to *dry fallout.*

Fatty acid — A carboxylic acid derived from or contained in an animal or vegetable fat or oil and composed of a chain of alkyl groups containing from 4 to 22 carbon atoms and characterized by a terminal carboxyl radical $-COOH$.

Fatty alcohol — A primary alcohol, composed usually of a long, straight-chain, organic molecule with 8 to 20 carbon atoms, terminating with an hydroxyl radical $-OH$. Fatty alcohols are oily liquids from C_8 to C_{11} and solids above C_{11}.

Fatty ester — A fatty acid with the active hydrogen replaced by the alkyl group of a monohydric alcohol. The esterification of a fatty acid, RCOOH, by an alcohol, ROH, yields the fatty ester RCOOR.

Fick's First Law — States that the rate of change in concentration of a substance is proportional to the second derivative of the concentration gradient. The proportionality constant is usually known as the diffusion coefficient.

Flux — Rate of movement of a substance across a surface of a given area in unit time.

Freon — A trade name for halocarbon, usually dichlorodifluoromethane, used for household pressurized dispensers and refrigeration systems.

Fungicide — A chemical substance used to destroy or control the growth of fungi.

Geochemical cycle — All the natural processes involved in the mobilization of materials in the earth's crust and ultimate transport into the oceans, with return of some of it (e.g., water and salts) to land via the atmosphere.

Glycoprotein — Plant or animal derived organic substance composed of various hydroxy compounds, sugars, and protein.

Henry's Law — States that the partial pressure of a gas in equilibrium with a solution is equal to a constant times its concentration in the solution; or when a liquid and a gas remain in contact, the weight of the gas that dissolves in a given quantity of liquid is proportional to the pressure of the gas above the liquid.

Intercalibration — An exercise to compare results of different laboratories applying similar or unlike methodologies for a particular analysis.

Interface — The boundary between different media (e.g., air-sea or sea-bottom interface) or between different densities of the same medium, as the interface between a freshwater layer overlying a seawater layer.

Laminar — Pertaining to or occurring in layers.

Lignite — A low form of coal between peat and subbituminous coal. Brown coal is a form of lignite that has not yet fully matured through all geologic processes that eventually form black coal.

Metabolite — A product of the metabolic process of plants or animals.

Metalloid — Metal-like element, e.g., arsenic, germanium.

Metallo-organic compound — The chemical or biochemical combination of a metal with an organic molecule, e.g., methyl mercury, tetraethyl lead.

Methylate — The chemical attachment of one or more methyl groups (CH_3) to a metal or other groups, e.g., dimethyl mercury.

Microlayer — The thin layer on the water surface, generally considered less than 100 μm thick, which has rather unique physical and chemical characteristics. The thickness reported in the literature varies according to the type of sampling equipment used operationally.

Nascent — Elemental form of a substance, e.g., oxygen and chlorine, having its full oxidizing or reducing power. Descriptive of the abnormally active condition of an element when it is released from chemical combination as, for example, nascent oxygen.

Oleophilic — Having an affinity for or being soluble in oil or a fatty substance.

Pathogen — A microorganism having the potential to cause disease.

Petroleum hydrocarbon — A compound or mixture of compounds composed of hydrogen and carbon and normally derived from subterranean sources produced by marine organisms through geological time.

Phagocytosis — Process by which living cells, e.g., white blood cells (leucocytes), ingest or engulf other cells or particles, such as bacteria and protozoa.

Proteoglycan — An organic compound derived from animals with protein attached to glycogen, the form in which carbohydrate is stored in animals.

Rafted bubbles — A collection of bubbles, visible as foam or froth, covering a finite area of the water surface.

"Red Tide" — A colloquial term for a large bloom of dinoflagellates, which give the sea surface a distinct reddish tinge and which may lead to paralytic shellfish poisoning, e.g., *Gonyaulax* sp.

Radionuclide (or nuclide) — Radioactive isotope of one of the chemical elements.

Rainout — The process of sorption (absorption and adsorption) of dissolved and particulate substances into a raindrop as it is initially formed in the cloud.

Sedimentation — The process of settling out of solid particulate material by gravitation in the atmosphere or the sea.

Seston — All particulate material, organic and inorganic, suspended in water.

Spore — The inactive, resting stage of an organism.

Stratosphere — That part of the earth's atmosphere between the tropopause (10 to 15 km altitude) and the stratopause (50 km altitude), where conditions are unaffected by the earth's weather.

Surface-active — The property of a substance which reduces surface tension and allows dissolution of one immiscible substance into another, e.g., soap when applied to oil and water.

Surfactant — A substance that reduces surface tension when dissolved in water or aqueous solution, or which reduces interfacial tension between two liquids or between a liquid and a solid. There are three categories of surfactants: detergents, wetting agents, and emulsifiers.

Transfer velocity — The rate with which a gas is transferred through the gas phase or the aqueous phase. It is defined as the quotient of the molecular diffusion coefficient divided by the layer thickness.

Transpiration — The emission of gases or vapor through excretory organs or pores, as in loss of moisture from plants.

Troposphere — The atmospheric envelope surrounding the earth and containing the earth's weather to an altitude of 10 to 15 km, where a discontinuity (the tropopause) separates it from the stratosphere.

Turbulence — The random mixing process in the atmosphere and the sea generated by winds, tides, and currents.

Vapor pressure — The pressure exerted in a closed system by a substance owing to its evaporation, measured usually under standard conditions of temperature and atmospheric pressure.

Washout — The process of removal of gaseous, vaporized, and particulate material from the atmosphere by raindrops in their descent from the clouds to the land or sea surface.

Whitecap — The visible white froth associated with breaking wave crests, which usually occurs at wind speeds exceeding 3 m sec^{-1}.

REFERENCES

1. Risebrough, R. W., Huggett, R. J., Griffin, J. J., and Goldberg, E. D., Pesticides: transatlantic movements in the northeast trades, *Science,* 159, 1233, 1968.
2. Bidleman, T. F. and Olney, C. E., Chlorinated hydrocarbons in the Sargasso Sea atmosphere and surface water, *Science,* 183, 516, 1974.
3. Bidlemen, T. F. and Olney, C. E., Long range transport of toxaphene insecticide in the atmosphere of the western North Atlantic, *Nature (London),* 257, 475, 1975.
4. NAS, *Radioactivity in the Marine Environment,* Panel on Radioactivity in the Marine Environment of the Committee on Oceanography, National Research Council, National Academy of Sciences, Washington, D.C., 1971, 272.
5. *Inadvertent Climate Modification,* Report of the Study of Man's Impact on Climate (SMIC), Massachusetts Institute of Technology Press, Cambridge, Mass., 1971, 308.
6. Machta, L., The role of the oceans and biosphere in the carbon dioxide cycle, in *The Changing Chemistry of the Oceans,* Dyrssen, D. and Jagner, D., Eds., Almqvist & Wiksell, Stockholm, 1972, 121.
7. Bolin, B., Changes in land biota and their importance for the carbon cycle, *Science,* 196, 613, 1977.
8. Goldberg, E. D., Ed., *The Nature of Seawater,* Dahlem Workshop Report, Dahlem Konferenzen, Abakon Verlagsgesellschaft, Berlin, 1975, 719.
9. NAS, *The Tropospheric Transport of Pollutants and other Substances to the Oceans,* Workshop on Tropospheric Transport of Pollutants to the Ocean Steering Committee, Ocean Sciences Board, Assembly of Mathematical and Physical Sciences, National Research Council, National Academy of Sciences, Washington, D.C., 1978, 243.
10. Williams, J., Ed., *Carbon Dioxide, Climate and Society,* Pergamon Press, Oxford, 1978, 332.
11. Overrein, L. N., A presentation of the Norwegian Project "Acid Precipitation — Effects on Forest and Fish," *Water Air Soil Pollut.,* 6, 167, 1976.
12. Wright, R. F., Dale, T., Gjessing, E. T., Hendrey, G. R., Henriksen, A., Johannessen, M., and Muniz, I. P., Impact of acid precipitation on freshwater ecosystems in Norway, *Water Air Soil Pollut.,* 6, 167, 1976.
13. Beamish, R. J., Loss of fish populations from unexploited remote lakes in Ontario, Canada as a consequence of atmospheric fallout of acid, *Water Res.,* 8, 85, 1974.
14. Beamish, R. J., Acidification of lakes in Canada by acid precipitation and the resulting effects on fishes, *Water Air Soil Pollut.,* 6, 501, 1976.
15. Beamish, R. J. and Van Loon, J. C., Precipitation loading of acid and heavy metals to a small acid lake near Sudbury, Ontario, *J. Fish. Res. Board Can.,* 34, 649, 1977.
16. Barker, D. R. and Zeitlin, H., Metal ion concentrations in sea-surface microlayer and size-separated atmospheric aerosol samples in Hawaii, *J. Geophys. Res.,* 77, 5076, 1972.
17. Duce, R. A., Hoffman, G. L., Ray, B. J., Fletcher, I. S., Wallace, G. T., Fasching, J. L., Piotrowitz, S. R., Walsh, P. R., Hoffman, E. J., Miller, J. M., and Heffter, J. F., Trace metals in the marine atmosphere: sources and fluxes, in *Marine Pollutant Transfer,* Windom, H. L. and Duce, R. A., Eds., D. C. Heath, Lexington, Mass., 1976, 77.
18. NAS, *Disposal in the Marine Environment: An Oceanographic Assessment,* An Analytical Study for the U.S. Environmental Protection Agency, Ocean Disposal Study Steering Committee, Commission on Natural Resources of the National Research Council, *National Academy of Sciences,* Washington, D.C., 1976, 76.
19. Baylor, E. R., Peters, V., and Baylor, M. B., Water-to-air transfer of virus, *Science,* 197, 763, 1977.
20. Baylor, E. R., Baylor, M. B., Blanchard, D. C., Syzdek, L. D., and Appel, C., Virus transfer from surf to wind, *Science,* 198, 575, 1977.
21. Gruft, H., Katz, J., and Blanchard, D. C., Postulated source of *Mycobacterium intracellulare* (BATTEY) infection, *Am. J. Epidemiol.,* 102, 311, 1975.
22. Windom, H. L. and Duce, R. A., Eds., *Marine Pollutant Transfer,* D. C. Heath, Lexington, Mass., 1976, 391.
23. MacIntyre, F., Chemical fractionation and sea surface microlayer properties, in *The Sea,* Vol. 5, Goldberg, E. D., Ed., John Wiley & Sons, New York, 1974, 245.
24. Liss, P. S., Chemistry of the sea surface microlayer, in *Chemical Oceanography,* Vol. 1, 2nd ed., Riley, J. P. and Skirrow, G., Eds., Academic Press, London, 1975, 193.
25. Hunter, K. A., Chemistry of the Sea Surface Microlayer, Ph.D. thesis, University of East Anglia, Norwich, England, 1977, 363.
26. Wangersky, P. J., The surface film as a physical environment, *Annu. Rev. Ecol. Syst.,* 7, 161, 1976.
27. Garrett, W. D., Collection of slick-forming materials from the sea surface, *Limnol. Oceanogr.,* 10, 602, 1965.
28. Harvey, G. W., Microlayer collection from the sea surface, *Limnol. Oceanogr.,* 11, 608, 1966.

29. Fasching, J. L., Courant, R. A., Duce, R. A., and Piotrowicz, S. R., A new surface microlayer sampler utilizing the bubble microtome, *J. Rech. Atmos.,* 8, 650, 1974.
30. University of Rhode Island, University of Connecticut, California Institute of Technology, Scripps Institute of Oceanography, CFR/CNRS, France, Yale University, Woods Hole Oceanographic Institution, Texas A & M University, and University of Miami, 1977 SEAREX, Sea-Air-Exchange, a proposal submitted to the National Science Foundation Office of the International Decade of Ocean Exploration, Washington, D.C., 1977, 56.
31. Dyer, A. J., A review of flux-profile relationships, *Boundary Layer Meteorol.,* 7, 363, 1974.
32. Duce, R. A., How does air pollution affect the oceans? *Maritimes,* 22, 4, 1978.
33. Garrett, W. D. and Smagin, V. M, Determination of the Atmospheric Contribution of Petroleum Hydrocarbons to the Oceans, World Meteorological Organization, Special Environmental Rept. No. 6 (WMO — No. 440), Geneva, 1976, 27.
34. Murozumi, M., Chow, T. J., and Patterson, C., Chemical concentrations of pollutant lead aerosols, terrestrial dusts, and sea salts in Greenland and Antarctic snow strata, *Geochim. Cosmochim. Acta,* 33, 1247, 1969.
35. Chesselet, R., Morelli, J., and Menard, P. B., Some aspects of the geochemistry of marine aerosols, in *The Changing Chemistry of the Oceans,* Dyrssen, D. and Jagner, D., Eds., Almqvist & Wiksell, Stockholm, 1972, 93.
36. Junge, C. E., *Air Chemistry and Radioactivity,* Academic Press, New York, 1963, 382.
37. Blanchard, D. C. and Syzdek, L. D., Variations in Aitken and giant-nuclei in marine air, *J. Phys. Oceanogr.,* 2, 255, 1972.
38. Woodcock, A. H., Salt nuclei in marine air as a function of altitude and wind force, *J. Meteorol.,* 10, 362, 1953.
39. Woodcock, A. H., Smaller salt particles in oceanic air and bubble behavior in the sea, *J. Geophys. Res.,* 77, 5316, 1972.
40. Woodcock, A. H., Blanchard, D. C., and Rooth, C. G. H., Salt-induced convection and clouds, *J. Atmos. Sci.,* 20, 159, 1963.
41. Sehnel, G. A. and Sutter, S. L., Particle deposition rates on water surface as a function of particle diameter and air velocity, *J. Rech. Atmos.,* 8, 911, 1974.
42. Junge, C. E., Our knowledge of the physico-chemistry of aerosols in the undisturbed marine environment, *J. Geophys. Res.,* 77, 5183, 1972.
43. Wang, C. S. and Street, R. L., Measurements of spray at an air-water interface, *Dyn. Atmos. Oceans,* 2, 141, 1978.
44. Szekielda, K.-H., Kupferman, S. L., Klemas, V., and Polis, D. F., Element enrichment in organic films and foam associated with aquatic frontal systems, *J. Geophys. Res.,* 77, 5278, 1972.
45. Sick, L. V., Johnson, C. C., and Engel, R., Trace metal enhancement in the biotic and abiotic components of an estuarine tidal front, *J. Geophys. Res.,* 83, 4659, 1978.
46. Pellenbarg, R. E. and Church, T. M., The estuarine surface microlayer and trace metal cycling in a salt marsh, *Science,* 203, 1010, 1979.
47. Garrett, W. D., Damping of capillary waves at the air-sea interface by oceanic surface-active material, *J. Mar. Res.,* 25, 279, 1967.
48. Jarvis, N. L., Garrett, W. D., Scheiman, M. A., and Timmons, C. O., Surface chemical characterization of surface active material in sea water, *Limnol. Oceanogr.,* 12, 88, 1967.
49. Baier, R. E., Goupil, D. W., Perlmutter, S., and King, R., Dominant chemical composition of sea-surface films, natural slicks, and foams, *J. Rech. Atmos.,* 8, 571, 1974.
50. Kattner, G. G. and Brockman, U. H., Fatty acid composition of dissolved and particulate matter in surface films, *Mar. Chem.,* 6, 233, 1978.
51. Slinn, W. G. N., Hasse, L., Hicks, B. B., Hogan, A. W., Lal, D., Liss, P. S., Munnich, K. O., Sehnel, G. A., and Vitteri, O., Some aspects of the transfer of atmospheric trace constituents past the air-sea interface, *Atmos. Environ.,* 12, 2055, 1978.
52. Brtko, W. J. and Kabel, R. L., Transfer of gases at natural air-water interfaces, *J. Phys. Oceanogr.,* 8, 543, 1978.
53. Liss, P. S. and Slater, P. G., Flux of gases across the air-sea interface, *Nature (London),* 247, 181, 1976.
54. Broecker, W. S. and Peng, T. H., Gas exchange rates between air and sea, *Tellus,* 26, 21, 1975.
55. Harvey, G. W. and Burzell, L. A., A simple microlayer method for small samplers, *Limnol. Oceanogr.,* 17, 156, 1972.
56. Miget, R., Kator, H., Oppenheimer, C., Laseter, J. L., and Ledet, E. J., New sampling device for the recovery of petroleum hydrocarbons and fatty acids from aqueous surface films, *Anal. Chem.,* 46, 154, 1974.
57. Ledet, E. J. and Laseter, J. L., A comparison of two sampling devices for the recovery of organics from aqueous surface films, *Anal. Lett.,* 7, 553, 1974.

58. Larsson, K., Odham, G., and Sodergren, A., On lipid surface films on the sea. I. A simple method for sampling and studies of composition, *Mar. Chem.*, 2, 49, 1974.
59. Garrett, W. D. and Barger, W. R., Sampling and Determining the Concentration of Film-Forming Organic Constituents of the Air-Water Interface, Memorandum Rept. 2852, Naval Research Laboratory, Washington, D.C., 1974, 13.
60. Baier, R. E., Surface quality assessment of natural bodies of water, in *Proc. 13th Conf. Great Lakes Research,* International Association on Great Lakes Research, Buffalo, New York, 1970, 114.
61. Baier, R. E., Organic films on natural waters; their retrieval, identification and modes of elimination, *J. Geophys. Res.*, 77, 5062, 1972.
62. MacIntyre, F., Bubbles: a boundary layer "microtome" for micron thick samples of a liquid surface, *J. Phys. Chem.*, 72, 589, 1968.
63. Morris, R. J., Lipid composition of surface films and zooplankton from the Eastern Mediterranean, *Mar. Pollut. Bull.*, 5, 105, 1974.
64. Crow, S. A., Ahern, D. G., and Cook, W. L., Densities of bacteria and fungi in coastal surface films as determined by a membrane absorption procedure, *Limnol. Oceanogr.*, 20, 644, 1975.
65. Hamilton, E. I. and Clifton, R. J., Techniques for sampling the air-sea interface for estuarine coastal waters, *Limnol. Oceanogr.*, 24, 188, 1979.
66. Hatcher, R. F. and Parker, B. C., Laboratory comparison of four surface microlayer samplers, *Limnol. Oceanogr.*, 19, 162, 1974.
67. Garrett, W. D., Comments on laboratory comparison of four microlayer samplers, *Limnol. Oceanogr.*, 19, 166, 1974.
68. Hsu, Y.-H. L. and Wu, J., Simultaneous measurements of bubbles and spray produced by breaking waves, *Trans. Am. Geophys. Union*, 61(Abstr.), 262, 1980.
69. Blanchard, D. C., The electrification of the atmosphere by particles from bubbles in the sea, *Prog. Oceanogr.*, 1, 71, 1963.
70. MacIntyre, F., Ion Fractionation in Drops from Breaking Bubbles, Ph.D. thesis, Massachusetts Institute of Technology, Cambridge, 1965, 272.
71. Ross, D. B. and Cardone, V., Observations of oceanic whitecaps and their relation to remote measurements of surface wind speed, *J. Geophys. Res.*, 79, 444, 1974.
72. Mason, B. J., The oceans as a source of cloud-forming nuclei, *Geofis. Pura Appl.*, 36, 148, 1957.
73. Blanchard, D. C., The oceanic production rate of cloud nuclei, *J. Rech. Atmos.*, 4, 1, 1969.
74. Blanchard, D. C. and Woodcock, A. H., Bubble formation and modification in the sea and its meteorological significance, *Tellus*, 9, 145, 1957.
75. Blanchard, D. C., Sea-to-air transport of surface active material, *Science*, 146, 396, 1964.
76. Blanchard, D. C. and Syzdek, L., Mechanism for the water-to-air transfer and concentration of bacteria, *Science*, 170, 626, 1970.
77. Blanchard, D. C., Whitecaps at sea, *J. Atmos. Sci.*, 28, 645, 1971.
78. Blanchard, D. C. and Syzdek, L. D., Concentration of bacteria from bursting bubbles, *J. Geophys. Res.*, 77, 5087, 1972.
79. Blanchard, D. C., International Symposium on the Chemistry of Sea/Air Particulate Exchange Processes: Summary and Recommendations, *J. Rech. Atmos.*, 8, 509, 1974.
80. Blanchard, D. C. and Syzdek, L. D., Importance of bubble scavanging in the water-to-air transfer of organic material and bacteria, *J. Rech. Atmos.*, 8, 529, 1974.
81. Blanchard, D. C. and Syzdek, L. D., Bubble tube: apparatus for determining rate of collection of bacteria by an air bubble rising in water, *Limnol. Oceanogr.*, 19, 133, 1974.
82. Blanchard, D. C., Bubble scavenging and the water-to-air transfer of organic material in the sea, in *Applied Chemistry at Protein Interfaces,* Baier, R., Ed., American Chemical Society, Washington, D.C., 1975, 360.
83. Blanchard, D. C. and Syzdek, L. D., Electrostatic collection of jet and film drops, *Limnol. Oceanogr.*, 20, 762, 1975.
84. Blanchard, D. C., Jet drop enrichment of bacteria, virus, and dissolved organic material, *Pageoph,* 116, 320, 1978.
85. Blanchard, D. C. and Syzdek, L. D., Seven problems in bubble and jet drop researches, *Limnol. Oceanogr.*, 23, 389, 1978.
86. Blanchard, D. C. and Hoffman, E. J., Control of jet drop dynamics by organic material in seawater, *J. Geophys. Res.*, 83, 6187, 1978.
87. MacIntyre, F., Enhancement of gas transfer by interfacial ripples, *Phys. Fluids*, 14, 1596, 1971.
88. MacIntyre, F., Additional problems in bubble and jet drop research, *Limnol. Oceanogr.*, 23, 571, 1978.
89. Green, T. and Houk, D. F., The removal of organic surface films by rain, *Limnol. Oceanogr.*, 24, 966, 1979.

90. Beauford, W. J., Barker, J., and Barringer, A. R., Release of particles containing metals from vegetation into the atmosphere, *Science,* 195, 571, 1977.

91. Kozuchowski, J. and Johnson, D. L., Gaseous emissions of mercury from an aquatic vascular plant, *Nature (London),* 274, 468, 1978.

92. Rasmussen, R. A., Emission of biogenic H_2S, *Tellus,* 26, 254, 1974.

93. Cattell, F. C. R. and Scott, W. D., Copper in aerosol particles produced by the ocean, *Science,* 202, 429, 1978.

94. Baier, R. E., Baylor, E. R., Blanchard, D. C., Carlucci, A. F., Dimmick, R. L., Duce, R. A., Hardy, C. D., and Woodcock, A. H., Report of the Panel on the Sea-to-Air Transfer of Material in the Sea, National Research Council Ocean Dumping Workshop, Woods Hole, Mass., Sept. 9 to 13, 1974, 31.

95. Zobell, C. E., Microorganisms in marine air, *Publ. Am. Assoc. Adv. Sci.,* 17, 55, 1942.

96. Woodcock, A. H., Note concerning human respiratory irritation associated with high concentrations of plankton and mass mortality of marine organisms, *J. Mar. Res.,* 7, 56, 1948.

97. Spendlove, J. C., Industrial, agriculture, and microbial aerosol problems, in *Developments in Industrial Microbiology,* Halvorson, New York, 1974, 20.

98. Cornwell, J., Is the Mediterranean dying? *The New York Times Magazine,* February 21, 1971.

99. CEQ, *Fluorocarbons and the Environment,* Report of the Federal Task Force on Inadvertent Modification of the Stratosphere (IMOS), Council on Environmental Quality, Federal Council for Science and Technology, Washington, D.C., 1975, 109.

100. NAS, *Halocarbons: Effects on Stratospheric Ozone,* National Academy of Sciences, Washington, D.C., 1976, 352.

101. Bidleman, T. F. and Olney, C. E., High volume collection of atmospheric PCB, *Bull. Environ. Contam. Toxicol.,* 11, 442, 1974.

102. Hasse, L. and Liss, P. S., Gas Exchange Across the Air-Sea Interface, Report to the Second Session of the GESAMP Working Group on Interchange of Pollutants between the Atmosphere and the Oceans, West Vancouver, B.C., September 1978.

103. Lovelock, J. E., Maggs, R. J., and Wade, R. J., Halogenated hydrocarbons in and over the Atlantic, *Nature (London),* 241, 194, 1973.

104. Bidleman, T. F., Rice, C. P., and Olney, C. E., High molecular weight chlorinated hydrocarbons in the air and sea: rates and mechanisms of air/sea transfer, in *Marine Pollutant Transfer,* Windom, H. L. and Duce, R. A., Eds., D. C. Heath, Lexington, Mass., 1976, 323.

105. Meszaros, E., Concentration of sulfur compounds in remote continental and oceanic areas, *Atmos. Environ.,* 12, 699, 1978.

106. NAS, *Petroleum in the Marine Environment,* Report on the Workshop on Inputs, Fates and Effects of Petroleum in the Marine Environment, National Academy of Sciences, Washington, D.C., 1975, 107.

107. Blumer, M., Submarine seeps: are they a major source of open ocean oil pollution?, *Science,* 176, 1257, 1972.

108. Wilson, R. D., Monaghan, P. H., Osanik, A., Price, L. M., and Rogers, M. A., National marine oil seepage, *Science,* 184, 857, 1974.

109. Harvey, G. R., Requejo, A. G., McGillivary, P. A., and Tokar, J. M., Observations of a subsurface oil-rich layer in the open ocean, *Science,* 205, 999, 1979.

110. Duce, R. A., Atmospheric hydrocarbons and their relation to marine pollution, in *Background Papers for a Workshop on Inputs, Fates, and Effects of Petroleum in the Marine Environment,* Vol. 2, National Academy of Sciences, Washington, D.C., 1973, 416.

111. Duce, R. A., Speculations on the budget of particulate and vapor phase non-methane organic carbon in the global troposphere, *Pageoph,* 116, 244, 1978.

112. Goldberg, E. D., Ed., *A Guide to Marine Pollution,* Gordon and Breach, New York, 1972, 168.

113. Irukayama, K., The pollution of Minamata Bay and Minamata disease, *Adv. Water Pollut. Res.,* 3, 153, 1967.

114. Goldberg, E. D., Rock volatility and aerosol composition, *Nature (London),* 260, 128, 1976.

115. Desaedeleer, G. and Goldberg, E. D., Rock volatility — some initial experiments, *Geochem J.,* 12, 75, 1978.

116. Brimblecombe, R. and Hunter, K. A., Rock volatility and aerosol composition, *Nature (London),* 265, 761, 1977.

117. Weiss, H. V., Koide, M., and Goldberg, E. D., Mercury in a Greenland ice sheet: Evidence of recent input by man, *Science,* 174, 692, 1971.

118. Hobbs, R. V., Radke, L. F., and Stith, J. L., Eruptions of the St. Augustine volcano: airborne measurements and observations, *Science,* 195, 871, 1977.

119. Mroz, E. and Zoller, W. H., Composition of atmospheric particulate matter from the eruption of Heimaey, Iceland, *Science,* 190, 461, 1975.

120. **Duce, R. A., Quinn, J. G., Olney, C. E., Piotrowicz, S. R., Roy, B. J., and Wade, T. L.,** Enrichment of heavy metals and organic compounds in the surface microlayer of Narragansett Bay, Rhode Island, *Science,* 176, 161, 1972.

121. **Piotrowicz, S. R., Ray, B. J., Hoffman, G. L., and Duce, R. A.,** Trace metal enrichment in the sea surface microlayer, *J. Geophys, Res.,* 77, 5243, 1972.

122. **Van Grieken, R. E., Johansson, T. B., and Winchester, J. W.,** Trace metal fractionation effects between sea water and aerosols from bubble bursting, *J. Rech. Atmos.,* 8, 611, 1974.

123. **Owen, R. M., Meyers, P. A., and Machin, J. E.,** Influence of physical processes on the concentration of heavy metals and organic carbon in the surface microlayer, *Geophys. Res. Lett.,* 6, 147, 1979.

124. **Piotrowicz, S. R., Duce, R. A., Fasching, J. L., and Waisel, C. P.,** Bursting bubbles and their effect on the sea-to-air transport of Fe, Cu and Zn, *Mar. Chem.,* 7, 307, 1979.

125. **Jernelöv, A.,** Mercury — a case study of marine pollution, in *The Changing Chemistry of the Oceans,* Dyrssen, D. and Jagner, D., Eds., Almqvist & Wiksell, Stockholm, 1972, 101.

126. **Wood, J. M.,** Biological cycles for toxic elements in the environment, *Science,* 183, 1049, 1974.

127. **Wong, P. T. S., Chau, Y. K., and Luxon, P. L.,** Methylation of lead in the environment, *Nature (London),* 253, 263, 1975.

128. **Schmidt, U. and Huber, F.,** Methylation of organolead and lead (II) compounds to $(Ch_3)_4$ Pb by microorganisms, *Nature (London),* 259, 157, 1976.

129. **Goldberg, E. D.,** Atmospheric dust, the sedimentary cycle and man, *Comments Geophys. Earth Sci.,* 1, 117, 1971.

130. **Eriksson, E.,** The yearly circulation of chloride and sulfur in nature; meteorological, geochemical and pedological implications. I, *Tellus,* 11, 375, 1959.

131. **Taylor, S. R.,** Abundance of chemical elements in the continental crust: a new table, *Geochim. Cosmochim. Acta,* 28, 1273, 1964.

132. **Riley, J. P.,** Analytical chemistry of sea water, in *Chemical Oceanography,* 2nd ed., Riley, J. P. and Skirrow, G., Eds., Academic Press, London, 1975, 193.

133. **Chester, R. and Stoner, S. H.,** The distribution of Mn, Fe, Cu, Ni, Co, Ga, Cr, V, Ba, Sr, Sn, Zn, and Pb in some soil-sized particulates from the lower troposphere over the world ocean, *Mar. Chem.,* 2, 157, 1974.

134. **Bertine, K. K. and Goldberg, E. D.,** Fossil fuel combustion and the major sedimentary cycle, *Science,* 173, 233, 1971.

135. **Patterson, C., Settle, D., Schaule, B., and Burnett, M.,** Transport of pollutant lead to the oceans and within ocean ecosystems, in *Marine Pollutant Transfer,* Windom, H. L. and Duce, R. A., Eds., D. C. Heath, Lexington, Mass., 1976, 23.

136. **Maenhaut, W., Zoller, W. H., Duce, R. A., and Hoffman, G. L.,** Concentration and size distribution of particulate trace elements in the South Polar atmosphere, *J. Geophys. Res.,* 84(C5), 2421, 1979.

137. **Zoller, W. H., Gladney, E. S., and Duce, R. A.,** Atmospheric concentrations and sources of trace metals at the South Pole, *Science,* 183, 198, 1974.

138. **Duce, R. A., Parker, P. L., and Giam, C. S.,** *Pollutant Transfer to the Marine Environment,* Deliberations and Recommendations of the NSF/IDOE Pollutant Transfer Workshop, 1974, University of Rhode Island, Kingston, 1974, 55.

139. **Goldberg, E. D.,** *The Health of the Oceans,* The Unesco Press, Paris, 1976, 172.

140. **Goldberg, E. D.,** Man's role in the major sedimentary cycle, in *The Changing Chemistry of the Oceans,* Dyrssen, D. and Jagner, D., Eds., Almqvist & Wiksell, Stockholm, 1972, 267.

141. **Skinner, B. J. and Turekian, K. K.,** *Man and the Ocean,* Prentice-Hall, Englewood Cliffs, N.J., 1973, 149.

142. **Duce, R. A. and Hoffman, G. L.,** Atmospheric vanadium transport to the ocean, *Atmos. Environ.,* 10, 989, 1976.

143. **WMO,** Report of the WMO Air Pollution Measurement Techniques Conf. 1976, World Meteorological Organization, Geneva, 1976, p. 4 and Annexes I—IV.

144. **Volchok, H. L., Bowen, V. T., Folsom, T. R., Broecker, W. S., Schuert, E. A., and Bien, G. S.,** Oceanic distributions of radionuclides from nuclear explosions, in *Radioactivity in the Marine Environment,* National Academy of Sciences, Washington, D.C., 1971, 42.

145. **Joseph, A. D., Gustafson, P. F., Russell, I. R., Schuert, E. A., Volchok, H. L., and Tamplin, A.,** Sources of radioactivity and their characteristics, in *Radioactivity in the Marine Environment,* National Academy of Sciences, Washington, D.C., 1971, 6.

146. **WHO,** *Health Implications of Nuclear Power Production,* World Health Organization Regional Publications, European Ser., No. 3, Copenhagen, 1978, 75.

147. **Tymen, G.,** Distribution of natural radioactivity attached to atmospheric aerosol particles, *J. Environ. Sci. Health,* 13, 803, 1978.

148. **Wong, K. M., Hodge, V. F., and Folsom, T. R.,** Plutonium and polonium inside giant brown algae, *Nature (London),* 237, 460, 1972.

149. **Folson, T. R., Hodge, V. F., and Gurney, M. E.,** Plutonium observed on algal surfaces in the ocean, *Mar. Sci. Commun.,* 1, 39, 1975.
150. **Machta, L., Hanson, K., and Keeling, C. D.,** Atmospheric Carbon Dioxide and Some Interpretation, Office of Naval Research Conf. Fate of the Fossil Fuel Carbonates, Honolulu, January 19 to 23, 1976.
151. **Bolin, B.,** The carbon cycle, *Sci. Am.,* September 1970, 125.
152. **Baes, C. F., Jr., Goeller, H. E., Olson, J. S., and Rotty, R. M.,** Carbon dioxide and climate: the uncontrolled experiment, *Am. Sci.,* 65, 310, 1977.
153. **Stewart, R. W.,** The oceans, the climate and people (with a view from Mars), *Atmos. Ocean,* 16, 367, 1978.
154. **Liss, P. S.,** Processes of gas exchange across an air-water interface, *Deep Sea Res.,* 20, 221, 1973.
155. **UNEP,** *The Environmental Impacts of Production and Use of Energy,* I. Fossil Fuels, United Nations Environment Programme, Report of the Executive Director, Energy Rept. Ser. ERS-1-79, Nairobi, Kenya, 1979, 97.
156. **Knabe, W.,** Effects of sulfur dioxide on terrestrial vegetation, *Ambio,* 5, 213, 1976.
157. **Hendry, G. R., Baalsrud, K., Traaen, T. S., Laake, M., and Raddum, G.,** Acid precipitation: some hydrobiological changes, *Ambio,* 5, 224, 1976.
158. **Ottar, B.,** Monitoring long-range transport of air pollutants: the OECD Study, *Ambio,* 5, 203, 1976.
159. **Oden, S.,** The acidity problem — an outline of concepts, *Water Air Soil Pollut.,* 6, 137, 1976.
160. **Van Loon, J. C. and Beamish, R. J.,** Heavy-metal contamination by atmospheric fallout of several Flin Flon area lakes and the relation to fish populations, *J. Fish. Res. Board Can.,* 34, 899, 1977.
161. **Beamish, R. J. and Harvey, H. H.,** Acidification of the La Cloche Mountain Lakes, Ontario, and resulting fish mortalities, *J. Fish Res. Board Can.,* 29, 1131, 1972.
162. **Beamish, R. J.,** Growth and survival of white suckers (*Catostomus commersoni*) in an acidified lake, *J. Fish. Res. Board Can.,* 31, 49, 1974.
163. **Coghill, C. V.,** The history and character of acid precipitation in eastern North America, *Water Air Soil Pollut.,* 6, 407, 1976.
164. **Matheson, D. H. and Elder, F. C., Eds.,** Atmospheric Contribution to the Chemistry of Lake Waters, First Specialty Symposium of the International Association for Great Lakes Research, Longford Mills, Ontario, Canada, 28 Sept. — 1 Oct. 1975, *J. Great Lakes Res.,* 2 (Suppl. 1), 1, 1976.
165. **Schofield, C. L.,** Water quality in relation to survival of brook trout, *Salvelinus fontinalis* (Mitchell), *Trans. Am. Fish. Soc.,* 94, 227, 1965.
166. **Schofield, C. L.,** Acid precipation: effects on fish, *Ambio,* 5, 228, 1976.
167. **Whelpdale, D. M. and Summers, P. W.,** Acid Precipitation in Canada, Rept. no. ARQT-5-75, Department of the Environment, Atmospheric Environment Service, Downsview, Ontario, 1975, 39.
168. **Likens, G. E., Wright, R. F., Galloway, J. N., and Butler, T. J.,** Acid Rain, *Sci. Am.,* 241, 43, 1979.
169. **Rasmussen, R. A.,** Emission of biogenic H_2S, *Tellus,* 26, 254, 1974.
170. **Hansen, M. H., Ingvorsen, K., and Jorgensen, B. B.,** Mechanisms of hydrogen sulfide release from coastal marine sediments to the atmosphere, *Limnol. Oceanogr.,* 23, 68, 1979.
171. **Aneja, V. P., Overton, J. H., Jr., Culpitt, L. T., Durham, J. L., and Wilson, W. E.,** Direct measurements of emission rates of some atmospheric biogenic sulfur compounds, *Tellus,* 31, 174, 1979.
172. **Adams, D. F., Farwell, S. O., Pack, M. R., and Bamesberger, W. L.,** Preliminary measurements of biogenic sulfur-containing gas emissions from soils, *J. Air Pollut. Contr. Assoc.,* 29, 380, 1979.
173. **Laird, A. R. and Miksad, R. W.,** Observations on the particulate chlorine distribution in the Houston-Galveston area, *Atmos. Environ.,* 12, 1537, 1978.
174. **Harbo, R. M., McComas, F. T., and Thompson, J. A. J.,** Fluoride concentrations in two Pacific coast inlets — an indication of industrial contamination, *J. Fish. Res. Board Can.,* 31, 1151, 1974.
175. **Enhalt, D. H.,** The CH_4 concentration over the ocean and its possible variation with latitude, *Tellus,* 30, 169, 1978.
176. **Fabian, P., Borchers, R., Weiler, K. H., Schmidt, U., Volz, A., Enhalt, D. H., Seiler, W., and Muller, F.,** Simultaneously measured vertical profiles of H_2, CH_4, CO, N_2O, $CFCl_3$ and CF_2Cl_2 in the mid-latitude stratosphere and troposphere, *J. Geophys. Res.,* 84, 3149, 1979.
177. **Bezdek, H. F. and Carlucci, A. F.,** Surface concentration of marine bacteria, *Limnol. Oceanogr.,* 17, 566, 1972.
178. **Carlucci, A. F. and Bezdek, H. F.,** On the effectiveness of a bubble for scavenging bacteria from seawater, *J. Geophys. Res.,* 77, 6608, 1972.
179. **Patterson, C. and Settle, D.,** Contribution of lead via aerosol deposition to the Southern California Bight, *J. Rech. Atmos.,* 8, 957, 1974.
180. **Chow, T. J. and Patterson, C. C.,** Concentration of barium and lead in Atlantic waters off Bermuda, *Earth Planet. Sci. Lett.,* 1, 397, 1966.
181. **Pattenden, N. J. and Goodman, G. T.,** personal communication, 1978.

INDEX